Managing Reality

Book Three

Managing the Contract

Bronwyn Mitchell and Barry Trebes

Published by Thomas Telford Publishing, Thomas Telford Ltd, 1 Heron Quay, London E14 4JD. URL: www.thomastelford.com

Distributors for Thomas Telford books are
USA: ASCE Press, 1801 Alexander Bell Drive, Reston, VA 20191-4400, USA
Japan: Maruzen Co. Ltd, Book Department, 3–10 Nihonbashi 2-chome, Chuo-ku, Tokyo 103
Australia: DA Books and Journals, 648 Whitehorse Road, Mitcham 3132, Victoria

First published 2005

Also available in this series from Thomas Telford Books
NEC – Managing Reality: Introduction to the Engineering and Construction Contract. ISBN 07277 3392 3
NEC – Managing Reality: Procuring an Engineering and Construction Contract. ISBN 07277 3393 1
NEC – Managing Reality: Managing change. ISBN 07277 3395 8
NEC – Managing Reality: Managing procedures. ISBN 07277 3396 6
NEC – Managing Reality: Complete box set. ISBN 07277 3397 4

Also available from Thomas Telford Books
NEC3 (complete box set). ISBN 07277 3382 6

A catalogue record for this book is available from the British Library

9 8 7 6 5 4 3 2 1

ISBN: 0 7277 3394 X

© Thomas Telford Limited 2005

Typeset by Academic + Technical, Bristol
Printed and bound in Great Britain by Bell & Bain Limited, Glasgow, UK

Preface

Now more than a decade on from its initial formal introduction (1st edition 1993), the NEC form of contract remains radical in its ethos and contemporary in its management principles for delivering successful projects in the business environment of 21st century construction. Although conceived in the mid-1980s, in what could be described as a decade of success and excess, with a construction industry racked by conflict and confrontation, the NEC owes much of its current widespread and growing usage to the deep recession of the early 1990s, which forced the construction industry to rethink its approach and performance, as much from a need to survive as from a desire to improve.

It is generally recognised and accepted that Sir Michael Latham's *Constructing the Team*, published in 1994, and Sir John Egan's *Rethinking Construction*, published in 1998, were the two main catalysts and energisers of change and improvement within the construction industry throughout the last decade of the 20th century and into the 21st century. Arguably the NEC was the third key driver of cultural reform and management discipline; indeed the NEC was formally recognised by Latham as being the contract form which, more than any other, aligned to his vision for future construction and Egan similarly embraced the NEC as part of his movement for innovation.

Usage has brought with it practical experience and accordingly the thrust of this book is about dealing with the reality of real-life projects. The authors have a combined knowledge and hands-on experience of NEC-managed projects spanning some 20 years and have therefore written the book with the specific purpose of advising and assisting those who would wish to use or even those who already do use the ECC, on the concepts of modern contract practices, procedures and administration.

This book is about 'how to': how to manage the ECC contract and how to administer it. As such, it does not attempt to give a legal treatise or a blow-by-blow review of each and every clause and certainly is not a rehash of the NEC/ECC Guidance Notes. It is intended to be complementary to other publications, which give excellent theoretical and legal perspectives. This book is about managing reality.

Although now regarded as a 'mature' form of contract the ECC is still relatively young when compared to more traditional forms. Therefore, even with current practitioners, experience will vary both in duration and depth. With this in mind the book has been consciously structured so as to be presented as a five-part book-set that covers the needs of the student professional or prospective client, through to the novice practitioner and experienced user. It provides a rounded view of the ECC, whatever your discipline, on both sides of the contractual relationship and is aimed at enabling everyone to realise the business benefits from using the NEC suite of contracts generally and the ECC in particular.

Since the NEC's official launch in 1993, adoption and usage of the NEC, renamed *Engineering and Construction Contract – ECC* in 1995 (2nd edition), has grown, such that it is now the most frequently used contract form for civils, transportation infrastructure and utilities works and is increasingly the preferred form for building construction projects. The latest edition of the contract (NEC3) was issued in July 2005.

The wide acceptance of the NEC generally and its elevation in stature as the preferred contract form is further reinforced by the Office of Government Commerce's endorsement of the third edition (NEC3). These five books take account of the changes introduced to the contract within NEC3.

Foreword

As almost the first UK client to use the NEC, even in its consultative form prior to its launch in 1993, I believe *NEC: Managing Reality* to be a welcome addition to the construction bookshelf: an essential introduction to the NEC for the prospective 'novice' practitioner and an excellent *aide-mémoire* reference book for the regular user of this form of contract.

The joy of this new book is that it brings together a wealth of practical expertise and knowledge from two of the UK's most experienced NEC practitioners written in a style that keeps faith with NEC principles of clarity and simplicity, while respecting differing levels of knowledge within its potential readership.

I have always believed that choosing the contract form is as much a business issue as a construction decision and that business success only results from good management. Certainly this book gives emphasis to the ethos of managing for success rather than the reactive debate of failure. It imparts knowledge, understanding and practical experience of the Contract in use and equally stresses the roles, responsibilities and discipline of the management procedures that apply to everyone. I particularly like the five-module format, in that it presents itself in readable, manageable chunks that can be readily digested or revisited whatever the reader's previous experience of its use.

This book educates, giving guidance and confidence to anyone dealing with real contract issues, following both the spirit as well as the letter of the Contract. It is arguably the most comprehensive practical treatise to date on how to manage and administer the ECC. It is a must for the professional office. Every 'home' should have one.

David H Williams, CEng, FICE
Chairman, Needlemans

(Formerly Group Construction and Engineering Director BAA plc;
founding Chairman, NEC UK Users Group: 1994–1997)

Introduction

General

This series of books will provide the people who are actually using the Engineering and Construction Contract (ECC) in particular, and the New Engineering Contract (NEC) suite in general, practical guidance as to how to prepare and manage an ECC contract with confidence and knowledge of the effects of their actions on the Contract and the other parties.

Each book in the series addresses a different area of the management of an ECC contract.

- Book One –   NEC Managing Reality: Introduction to the Engineering and Construction Contract
- Book Two –   NEC Managing Reality: Procuring an Engineering and Construction Contract
- Book Three – NEC Managing Reality: Managing the Contract
- Book Four –   NEC Managing Reality: Managing Change
- Book Five –   NEC Managing Reality: Managing Procedures

- *Book One (NEC Managing Reality: Introduction to the Engineering and Construction Contract)* is for those who are considering using the ECC but need further information, or those who are already using the ECC but need further insight into its rationale. It therefore focuses on the fundamental cultural changes and mind-shift that is required to successfully manage the practicalities of the ECC in use.

- *Book Two (NEC Managing Reality: Procuring an Engineering and Construction Contract)* is for those who need to know how to procure an ECC contract. It covers in practical detail the invitations to tender, evaluation of submissions, which option to select, how to complete the Contract Data and how to prepare the Works Information. The use of this guidance is appropriate for employers, contractors (including subcontractors) and construction professionals generally.

- *Book Three (NEC Managing Reality: Managing the Contract)* is essentially for those who use the contract on a daily basis, covering the detail of practical management such as paying the contractor, reviewing the programme, ensuring the quality of the works and dispute resolution. Both first-time and experienced practitioners will benefit from this book.

- *Book Four (NEC Managing Reality: Managing Change)* is for those who are managing change under the contract; whether for the employer or the contractor (or subcontractor) the management of change is often a major challenge whatever the form of contract. The ECC deals with change in a different way to other more traditional forms. This book sets out the steps to efficiently and effectively manage change, bridging the gap between theory and practice.

- *Book Five (NEC Managing Reality: Managing Procedures)* gives step-by-step guidance on how to apply the most commonly used procedures, detailing the actions needed by all parties to comply with the contract. Anyone administering the contract will benefit from this book.

Background

The ECC is the first of what could be termed a 'modern contract' in that it seeks to holistically align the setting up of a contract to match business needs as opposed to writing a contract that merely administers construction events.

The whole ethos of the ECC, or indeed the NEC suite generally, is one of simplicity of language and clarity of requirement. It is important that the roles and responsibilities are equally clear in definition and ownership.

When looking at the ECC for the first time it is very easy to believe that it is relatively straightforward and simple. However, this apparent simplicity belies the need for the people involved to think about their project and their role and how the ECC can deliver their particular contract strategy.

The ECC provides a structured flexible framework for setting up an appropriate form of contract whatever the selected procurement route. The fundamental requirements are as follows.

- The Works Information – quality and completeness – what are you asking the Contractor to do?
- The Site Information – what are the site conditions the Contractor will find?
- The Contract Data – key objectives for completion, for example start date, completion date, programme – when do you want it completed?

The details contained in the series of books will underline the relevance and importance of the above three fundamental requirements.

The structure of the books

Each chapter starts with a synopsis of what is included in that chapter. Throughout the book there are shaded 'practical tip' boxes that immediately point the user towards important reminders for using the ECC (see example below).

> Clarity and completeness of the Works Information is fundamental.

There are also unshaded boxes that include examples to illustrate the text (see example below).

> Imagine a situation in which the *Supervisor* notifies the *Contractor* that the reinstatement of carriageways on a utility diversion project is not to the highway authority's usual standards. However, the Works Information is silent about the reinstatement.
>
> Although it is not to the authority's usual standard, it is **not** a Defect because the test of a Defect is non-conformance with the Works Information. In this situation, if the *works* need to be redone to meet the authority's requirements, the *Contractor* is entitled to a compensation event because the new requirements are a change to the Works Information.

Other diagrams and tables are designed to maintain interest and provide another medium of explanation. There are also standard forms for use in the administration and management of the contract together with examples.

Throughout the books, the following terms have been used in a specific way.

- NEC is the abbreviation for the suite of New Engineering Contracts and it is not the name of any single contract.
- ECC is the abbreviation for the contract in the NEC suite called the Engineering and Construction Contract.

The NEC suite currently comprises the

- Engineering and Construction Contract
- Engineering and Construction Subcontract

- Engineering and Construction Short Contract
- Engineering and Construction Short Subcontract
- Professional Services Contract
- Adjudicator's Contract
- Term Service Contract
- Framework Contract

Acknowledgements

We would like to thank the following individuals and companies who have supported the book.

Andy Door, who gave advice on procurement within the public sector. Mike Attridge of Ellenbrook Consulting who reviewed the book on behalf of the authors. David H. Williams who provided guidance and support in the development of the book and everyone at Needlemans Construction Consultants and MPS Limited.

Series Contents

The following outlines the content of the five books in the series.

Book 1 NEC Managing Reality: Introduction to the Engineering and Construction Contract

Preface, Foreword, Introduction and Acknowledgements
Series Contents, Contents, List of tables, List of figures

Chapter 1 Introduction to the Engineering and Construction Contract, concepts and terminology

Synopsis
This chapter looks at:

- An introduction to the ECC
- An identification of some of the differences between the ECC and other contracts
- A brief outline of differences between ECC2 and ECC3
- An outline of the key features of the ECC
- Conventions of the ECC
- Concepts on which the ECC is based
- Terminology used in the ECC
- Terminology not used in the ECC
- How the ECC affects the way you work

Appendix 1 Summary of differences between ECC2 and ECC3

Chapter 2 Roles in the Engineering and Construction Contract

Synopsis
This chapter describes the roles adopted in the ECC including:

- How to designate a role
- Discussion of the roles described in the ECC
- Discussion of the project team
- How the ECC affects each of the roles

Appendix 2 List of duties

Book 2 NEC Managing Reality: Procuring an Engineering and Construction Contract

Preface, Foreword, Introduction and Acknowledgements
Series Contents, Contents, List of tables, List of figures

Chapter 1 Procurement

Synopsis
This chapter looks at the concept of procurement and contracting strategies and discusses:

- Procurement and contract strategy
- What tender documents to include in an ECC invitation to tender
- How to draft and compile a contract using the ECC
- Procurement scenarios that an employer could face and how to approach them

- What are framework agreements and how they could incorporate the ECC
- What is partnering and how it can be used with the ECC

Appendix 1   Assessing tenders
Appendix 2   ECC tender documentation

Chapter 2   Contract Options

Synopsis
This chapter looks at the Contract Options available within the ECC:

- ECC main and secondary Options
- Priced contracts
- Target contracts
- Cost-reimbursable contracts
- Choosing a main Option
- Choosing a secondary Option

Appendix 3   Audit Plan

Chapter 3   Completing the Contract Data

Synopsis
This chapter gives guidance on:

- How to choose a main Option
- How to choose secondary Options
- Choosing optional statements in the Contract Data
- Where to position the optional statements in the Contract Data
- How to complete each statement in the Contract Data

Chapter 4   Works Information guidelines

Synopsis
This chapter looks at the Works Information and Site Information:

- Providing the Works
- What should be included in the Works Information
- Separation of the Works and Site Information
- Structuring for the Works Information
- Interface management
- General rules in drafting the Works Information
- Site Information

Appendix 4   Works Information clauses

Book 3   NEC Managing Reality: Managing the Contract

Preface, Foreword, Introduction and Acknowledgements
Series Contents, Contents, List of tables, List of figures

Chapter 1   Payment procedures in the ECC

Synopsis
This chapter discusses the following:

- The payment procedure including:
  - When the *Contractor*'s application for payment is submitted
  - When assessments take place
  - When the payment certificate is issued
  - How invoicing is carried out
  - When payment takes place
- The effects of Option Y(UK)2 taking into account the Housing Grants, Construction and Regeneration Act 1996

Chapter 2   Control of time

Synopsis
This chapter discusses aspects relating to the *Contractor*'s programme including:

- The terminology used to describe the programme
- What the programme is
- The definition and purpose of the Accepted Programme
- How and when to submit programmes
- What the programme is used for
- What to include in the programme

Chapter 3 Control of quality

Synopsis
This chapter discusses:

- The quality framework embedded within the ECC
- The *Contractor*'s obligations
- Role of the *Employer*'s representatives
- Subcontracting
- Quality control

Chapter 4 Disputes and dispute resolution

Synopsis
This chapter:

- Emphasises the importance of early dispute resolution to the successful outcome of a contract
- Considers the common sources of dispute
- Considers how the ECC has been designed to reduce the incidence of disputes
- Examines how the ECC provides for the resolution of disputes
- Looks at the implications for the dispute resolution process as a result of the new Housing Grants, Construction and Regeneration Act 1996
- Looks at ECC3 changes in relation to adjudication.

Book 4 NEC Managing Reality: Managing Change

Preface, Foreword, Introduction and Acknowledgements
Series Contents, Contents, List of tables, List of figures

Chapter 1 Compensation Events

Synopsis
This chapter describes the following:

- The compensation events contained within the ECC
- Procedure for administering compensation events
- Roles played by the two main parties to the contract in relation to compensation events

Appendix 1 Compensation event procedures

Chapter 2 Schedule of Cost Components

Synopsis
This chapter discusses aspects relating to the Schedule of Cost Components including:

- When the Schedule of Cost Components is used
- How the SCC interacts with the payment clauses
- Actual Cost and Defined Cost
- The Fee
- The components of cost included under the Schedule of Cost Components
- Contract Data part two

Appendix 2 Section A: ECC2 example quotations
Section B: ECC3 example quotations
Appendix 3 Example people costs
Appendix 4 Preliminaries comparison
Appendix 5 *Contractor*'s and Subcontractor's share example

Book 5 NEC Managing Reality: Managing Procedures

Preface, Foreword, Introduction and Acknowledgements
Series Contents, Contents, List of tables, List of figures

Chapter 1 ECC Management: Procedures

Synopsis

This chapter brings together all the aspects discussed in previous chapters in Books 1 to 4, which form part of the series of books on NEC Managing Reality. This chapter provides the 'how to' part of the series. It introduces some example pro-formas for use on the contract.

For quick reference, this chapter may be read on its own. It does not, however, detail the reasons for carrying out the actions, or the clause numbers that should be referred to in order to verify the actions in accordance with the contract. These are described in detail in other chapters that form part of this series.

Contents

1   Payment procedures in the Engineering and Construction Contract   **1**

    Synopsis   1

    1.1   Introduction   2

    1.2   Payment procedure   2

         1.2.1   *Contractor*'s application for payment   2

         1.2.2   Assessment date   3

         1.2.3   Assessing the amount due   4

         1.2.4   Certification   7

         1.2.5   Invoices   7

         1.2.6   Payment   8

         1.2.7   Representation of the payment procedure in the ECC   8

    1.3   How the ECC3 includes the HGCR Act 1996   8

    1.4   The effects of Y(UK)2   10

         1.4.1   Periods for payment   10

         1.4.2   Payment procedure with Y(UK)2   12

         1.4.3   Withholding payment with Y(UK)2   12

    1.5   Payment procedure on Completion   12

    1.6   Payment procedure after Completion   13

    1.7   Payment procedure after the *defects date*   13

2   Control of Time   **15**

    Synopsis   15

    2.1   Introduction   16

    2.2   Terminology   16

         2.2.1   Contract date   16

         2.2.2   *Starting date*   16

         2.2.3   ECC2 *possession date(s)* – ECC3 *access date(s)*   16

         2.2.4   Completion   17

         2.2.5   Completion date   17

         2.2.6   A Key Date

                 (ECC3-specific)   17

         2.2.7   Secondary Option L – sectional completion   17

         2.2.8   Planned Completion   17

    2.3   What is the programme?   18

    2.4   Definition of the Accepted Programme   18

         2.4.1   When the programme is submitted for acceptance   18

    2.5   The purpose of the Accepted Programme   20

         2.5.1   What the Accepted Programme can be used for   20

    2.6   What is included in the programme?   21

         2.6.1   All programmes   21

         2.6.2   All programmes except the first programme   22

         2.6.3   Notes on the programme inclusions   22

    2.7   Submission of the first programme   31

    2.8   How often is the programme revised?   31

    2.9   Acceptance of the programme   32

         2.9.1   Reasons for non-acceptance   33

         2.9.2   Resubmission of an unaccepted programme   34

         2.9.3   Timing of the acceptance or non-acceptance   34

    2.10 Completion   35

         2.10.1   Take over and Completion   35

    2.11 Take over by the *Employer*   35

	2.11.1	How it happens	35
	2.11.2	What it means	35
2.12	Acceleration		36
2.13	Other aspects of programming in the ECC		36
	2.13.1	*Contractor*'s programme	36
	2.13.2	*Employer*'s programme	37
	2.13.3	Affecting cash flow	37
	2.13.4	Moving target	37
	2.13.5	What happens if the scope of the *works* is reduced or increased?	38
	2.13.6	What happens if the activities on the revised programme do not match the *activity schedule*?	38
	2.13.7	What happens if the *Contractor* fails to maintain an Accepted Programme?	38
	2.13.8	The importance of good site records	38

3 Control of quality **40**

	Synopsis		40
3.1	Introduction		41
3.2	The ECC and quality		41
3.3	The *Contractor*'s obligations		41
	3.3.1	General obligations	41
	3.3.2	Ambiguities and inconsistencies (discrepancies) in or between the contract documents	42
	3.3.3	*Contractor*'s design	43
	3.3.4	Supervision/employees	44
	3.3.5	Mode and method of construction	44
	3.3.6	Setting out	46
	3.3.7	Quality management systems	46
3.4	Role of the *Employer*'s representatives with respect to quality		47
	3.4.1	*Project Manager*	48
	3.4.2	*Supervisor*	48
	3.4.3	*Employer*	49
3.5	*Employer*'s supply		49
	3.5.1	Plant and Materials, facilities and services	49
	3.5.2	Other contractors	49
3.6	Subcontracting		50
3.7	Quality control		52
	3.7.1	General	52
	3.7.2	Access for the *Employer* and his representative	52
	3.7.3	Test and inspections	52
	3.7.4	Notifying and investigating Defects and additional testing	53
	3.7.5	Quality procedures	54
3.8	Defective work		55
	3.8.1	General	55
	3.8.2	Rejection	55
	3.8.3	Correction of Defects	55
	3.8.4	Defects liability period	56
	3.8.5	Concessions	57
3.9	Certification		57
3.10	Enforcement		59
	3.10.1	General	59
	3.10.2	Incentivisation through certification	59
	3.10.3	Removal of employees	59
	3.10.4	Correction of Defects by Others	59
	3.10.5	Low performance damages	60
	3.10.6	Termination of the *Contractor*'s employment	60
3.11	NEC 3rd Edition		61

4 Disputes and dispute resolution **63**

| | Synopsis | 63 |
| 4.1 | Introduction | 64 |

	4.2	How disputes arise	64
		4.2.1 Introduction	64
		4.2.2 Interpretation of documents	65
		4.2.3 Cost and time effect of disputes	70
	4.3	How the ECC seeks to reduce the incidence of disputes	71
		4.3.1 Early warning	71
		4.3.2 Valuing changes	71
		4.3.3 Clear division of function and responsibility	72
		4.3.4 Reducing disputes in the Works Information and Site Information	73
		4.3.5 Conclusion	74
	4.4	Dispute resolution under the ECC	75
		4.4.1 General	75
		4.4.2 Adjudication: pre-HGCR Act 1996	75
		4.4.3 Adjudication: post-HGCR Act 1996	77
		4.4.4 Adjudication in ECC3	77
		4.4.5 The *tribunal*	78
	4.5	Adjudication – general comments and observations	81
		4.5.1 'Star Chambers' and the like	82
		4.5.2 Good information and records – how the *Adjudicator* will judge the information	82
		4.5.3 The *Adjudicator*	83
	4.6	NEC 3rd Edition	84
Index			**89**

List of tables

Table 1.1   ECC3 and Option Y(UK)2   9

Table 4.1   ECC3 Option W2 dispute resolution procedure – compliance with HGCR
            Act 1996   79

List of tables

List of figures

Fig. 1.1.   Default time period in the contract   8
Fig. 1.2.   Payment procedures in the ECC   8
Fig. 1.3.   Times for payment in the ECC section 5   11
Fig. 1.4.   Times for payment in Y(UK)2   11
Fig. 1.5.   Times for payment in ECC3 clause 51.2 and Option Y(UK)2   12
Fig. 1.6.   Payment procedure with Y(UK)2   12
Fig. 1.7.   Withholding payment procedure   13
Fig. 2.1.   Traditional clause 14 programme   19
Fig. 2.2.   Possession/access dates   24
Fig. 2.3.   Assessment of time effects of change   29
Fig. 2.4.   Secant piled lift shaft box   34
Fig. 2.5.   Reducing and increasing the scope of the *works*   38
Fig. 3.1.   Control of quality in the ECC   41
Fig. 4.1.   Dispute escalation   82
Fig. 4.2.   NEC3 adjudication options   85

1 Payment procedures in the Engineering and Construction Contract

Synopsis

This chapter discusses the following:

> - The payment procedure including:
>
>   - when the *Contractor*'s application for payment is submitted
>
>   - when assessments take place
>
>   - when the payment certificate is issued
>
>   - how invoicing is carried out
>
>   - when payment takes place
>
> - The effects of option Y(UK)2 taking into account the Housing Grants, Construction and Regeneration Act 1996

1.1 Introduction

One of the biggest changes made by the ECC in terms of payment is that it is the *Project Manager*'s job to assess the amount due to the *Contractor*. The *Contractor* may submit an application for payment, but is not obliged to do so. The lack of an application for payment from a *Contractor* could hinder the *Project Manager* in his assessment of the amount due, particularly for contract Options C, D and E.

Another point to note is that there are no contractual invoicing procedures.

Lastly, the deductions made from the amount due are made by the *Project Manager* and are obliged to be made by him. There is no discretion on his part not to deduct retention from the amount due if a first programme has not been received and there is no discretion not to deduct delay damages if Completion is late. It is not the *Employer* who makes these deductions but the *Project Manager*.

It is worth noting here that the power for the *Project Manager* to act in this way may not align with the internal procedures of the *Employer*. Sometimes the employer or finance officer of a business will wish to retain these powers for himself. This needs to be addressed for the ECC to work effectively.

The allocation of these powers in the ECC is a reflection of the importance which the contract places on the *Project Manager* and it emphasises the significance of the selection procedures for a *Project Manager* and that he should have the power and authority to act as described in the contract.

1.2 Payment procedure

The following is a breakdown of the payment procedure as outlined in the *conditions of contract*. The choice of main Option will affect what appears in the application for payment, but will not necessarily affect the procedure.

1.2.1 *Contractor*'s application for payment

The *Contractor* submits his application for payment to the *Project Manager* on or before the assessment date.[1] This step is not obligatory and is possibly not required for Option A or even Option B. It becomes essential for Options C, D and E, however, where the *Project Manager* is likely to find it difficult to make his assessment without some submission by the *Contractor* showing his costs in accordance with the Schedule of Cost Components. In Options C, D and E, the *Contractor* is paid his Actual Cost[2] plus his Fee; that is, his interim payments are based on the Schedule of Cost Components and not on some immediately determinable pricing tool such as an *activity schedule* or *bill of quantities*. Without first seeing what the *Contractor* considers his Actual Cost[3] plus Fee to be, the *Project Manager* is likely to find it virtually impossible to assess payment.

> The *Contractor* is not obliged in the *conditions of contract* to submit an application for payment.

Many *Employers* add into the Works Information or the *conditions of contract* using secondary Option Z that the *Contractor* is required to submit an application within a certain period of time (such as five days) prior to the assessment date. This firms up the obligation and makes it quite clear that the *Employer* wants an application for payment prior to the *Project Manager* assessing the amount due. In any case, contractors may be used to submitting 'interim valuations' and such an action may even be required by their own internal procedures.

---

[1] Clause 50.4.
[2] ECC3: Defined Cost.
[3] ECC3: Defined Cost.

The *Contractor*'s assessment comprises[4]

> • the Price for Work Done to Date
> plus • other amounts to be paid to the *Contractor*
> less • amounts to be paid by or retained from the *Contractor*
> plus • VAT as appropriate.

Ideally the *Contractor* should show the Price for Work Done to Date as a cumulative amount, less previous payments, leaving a payment for the assessment date in question.

1.2.2 Assessment date Assessments take place at each assessment date.[5]

1.2.2.1 First assessment date The first assessment date is decided by the *Project Manager* to suit the procedures of the Parties and is not later than the *assessment interval* after the *starting date*.[6] There are three aspects to note about this clause.

(1) It is the *Project Manager* who decides the date of the first assessment, rather than the *conditions of contract* or the Parties.

(2) Although the *Project Manager* decides the date of the first assessment, it must suit the procedures of both the *Employer* and the *Contractor*. In order to do this, input is required from the *Contractor* regarding what date would suit his procedures. This could take place at the start-up meeting, or by other communication prior to the *starting date*. The other assessment dates could be independent of this first date (they occur at the end of each *assessment interval* and the *assessment interval* need not be related to the first assessment date). The occurrence of this first date need not affect later assessments.

(3) The first assessment date cannot be later than the *assessment interval* after the *starting date*. It is the *starting date* that is the important date here, not the start date on site or *possession dates*.[7] This facilitates payment for work done off site prior to work starting on site, such as manufacture or design.

1.2.2.2 Other assessment dates Assessment dates other than the first assessment date differ for ECC2 and for ECC3.

(1) In ECC2, assessment dates other than the first assessment date occur at the end of each *assessment interval* until Completion of the whole of the *works*.[8] Therefore, until Completion is reached (as decided by the *Project Manager*), assessment dates occur regularly. After Completion, assessments take place as further described in clause 50.1 (ECC2).

In ECC3, assessment dates other than the first assessment date occur at the end of each *assessment interval* until four weeks after the *Supervisor* issues the Defects Certificate.[9] This differs from ECC2 in that the time period during which regular assessment dates take place is much longer in ECC3. The Defects Certificate is issued at the later of the *defects date* and the end of the last *defect correction period*.[10] This means that in ECC3, assessment dates take place regularly, from Completion until the later of the *defects date* and the end of the last *defect correction period*. Whereas ECC2 specified the situations when assessment dates would take place during this period, thus focusing the *Project Manager*'s assessment efforts when required, ECC3 appears to require assessments to take place regularly during a period of approximately one year, even though there may have been no activity on the contract and an assessment is therefore not necessary.

[4]Clause 50.2.
[5]Clause 50.1.
[6]Clause 50.1.
[7]ECC3: *access dates*.
[8]Clause 50.1 (ECC2).
[9]Clause 50.1 (ECC3).
[10]ECC3: clause 43.3.

The *assessment interval* is identified in Contract Data part one and is the key to the length of time between interim valuations of the project. Note that the trigger is the **end** of the *assessment interval*.

Examples of *assessment interval*s are as follows:

(a) Four weeks
Because not all months are four weeks long, the end of a four-week period will creep earlier and earlier and it will not be long before the end of the four-week period is somewhere in the middle of a month. This is not really convenient for either party.

(b) One calendar month
The end of a calendar month is more convenient and is also more definable without resorting to calendars to determine when the date will fall. If even more preciseness is required, the Parties could agree to the last Friday of each month, or whatever is suitable to their procedures.

If, for example, the *assessment interval* is 'one calendar month' (as stated in Contract Data part one), then the assessment dates will be at the end of each calendar month. This could be the last working day of each calendar month or the last Sunday in the month, with earlier dates in the month of December.

(c) In accordance with the attached schedule of project dates
Some employers, recognising that the project procedures could correlate with the accounting department's requirements, compile a schedule of project dates showing:

- when the *Contractor* submits his application for payment,
- the assessment date,
- the date of the payment certificate,
- the date of invoice (see below),
- the date for payment.

This matrix of project dates makes it quite clear to all parties concerned when documents are required to be submitted.

(2) An assessment also takes place at Completion of the whole of the *works*. In other words, assessments take place up to Completion (and after Completion for ECC3), but this statement makes it clear that an assessment also takes place when Completion occurs. The *Project Manager* decides the date of Completion and would therefore be in a position also to do an assessment.

(3) **ECC2 only**: The *Project Manager* is required to carry out an assessment after Completion of the whole of the *works* but before the issuing of the Defects Certificate when:

- a previous amount due is corrected, either by the *Project Manager* or by the *Adjudicator* or *tribunal*,
- a payment is made late: the *Project Manager* is required to assess interest in each certificate himself rather than wait for the *Contractor* to claim the interest for late payment.

(4) **ECC2 only**: Lastly, the *Project Manager* assesses the amount due four weeks after the *Supervisor* issues the Defects Certificate. This is clearly required if Option P Retention has been chosen and the second half of the retention is required to be released after the issue of the Defects Certificate. Other circumstances when this assessment could be required are when the *Contractor* has not corrected a Defect and the *Project Manager* has assessed the cost of having the Defect corrected by other people and the *Contractor* is required to pay that amount.[11]

1.2.3 Assessing the amount due   At each assessment date as identified above, the *Project Manager* assesses the amount due.[12] The *Project Manager* assesses the amount due in the same

---

[11] Clause 45.1.
[12] Clause 50.1 and clause 50.2.

way that the *Contractor* would; that is, by assessing the cumulative Price for Work Done to Date and then the amount now due.

> The *Project Manager* assesses the amount due.

1.2.3.1 What is included in the assessment

The amount due assessed by the *Project Manager* is the Price for Work Done to Date plus other amounts to be paid by the *Contractor* less amounts to be paid by or retained from the *Contractor*.[13] Any value added tax or sales tax which the law requires the *Employer* to pay to the *Contractor* is included in the amount due. This is similar to the way in which a *Contractor* would be expected to submit an application for payment. The amount due can be summarised as follows.

	(1)	the Price for Work Done to Date
plus	(2)	other amounts to be paid to the *Contractor*
less	(3)	amounts to be paid by or retained from the *Contractor*
		subtotal
less	(4)	previous payments
		payment due for this application
plus	(5)	VAT
		total to pay

(1)   The Price for Work Done to Date is a defined term that depends on the main Option chosen as part of the *Employer*'s contract strategy. For example, in Option A, the Price for Work Done to Date is the total of the Prices for completed activities that are without Defects. The *Project Manager* should therefore take the main Option into account when assessing the amount due.[14]

(2)   Other amounts to be paid to the *Contractor* could relate to an advanced payment (ECC2 Option J; ECC3 Option X14), a bonus for early Completion (ECC2 Option Q; ECC3 Option X6), a correction of previous certificates, and interest due.

(3)   Amounts to be paid by or retained from the *Contractor* could include retention (ECC2 Option P; ECC3 Option X16), the repayment of an advancement (ECC2 Option J; ECC3 Option X14), delay damages (ECC2 Option R; ECC3 Option X7) or the retention of 25% of the Price for Work Done to Date for a late first programme.

(4)   Since the Price for Work Done to Date is the total to date, in order to calculate the amount due in the current assessment, previous payments should be deducted.

(5)   VAT is to be shown separately, but included in the amount due.

1.2.3.2 Programme

The *Project Manager* determines whether the *Contractor* has submitted a first[15] programme for acceptance which shows the information required by the contract.[16] If the *Project Manager* has not received a first programme, then a quarter[17] (25%) of the Price for Work Done to Date[18] is deducted from the assessment of the amount due. Note that the criterion is not that the *Project Manager* has accepted the programme, but that one has been submitted, showing all the information required by the contract and the Works Information.

> The *Contractor*'s cash flow could suffer if he does not submit a first programme for acceptance when he is required to do so.

Where a programme was requested and submitted with the tender, this programme may not show all the information required by the contract, especially if

---

[13] Clause 50.2.
[14] More about this in Chapter 2 of Book 2, Contract Options.
[15] This clause refers only to the first programme. Other incentives, such as the *Project Manager* making his own assessment as contained within clause 64.2, exist for the *Contractor* submitting later programmes.
[16] Clause 50.3.
[17] This is an obligation, not an option.
[18] Not 25% of the amount due.

that information was not available at the time of tender. It is unlikely, however, that the *Project Manager* would be able to use this lack of information as the reason for deducting the retention allowed since the lack of information was not, in this case, the *Contractor*'s fault. If, on the other hand, the *Contractor* was required to submit a first programme within a certain time period after the Contract Date, then the *Contractor* should ensure that it is submitted at the latest by the first assessment date.

1.2.3.3 Interest

Note that it is the *Project Manager* who assesses the interest and includes it in the certificate for payment. The *Project Manager* should not wait for the *Contractor* to claim the interest. Interest is paid for three reasons.

(1) If a payment was made late,[19] interest is assessed on the late payment and paid in the very next assessment.[20] The interest is assessed from the date when the payment should have been made to the date when it was made.

(2) If an amount due is corrected in a later certificate (whether by the *Project Manager*, the *Adjudicator* or the *tribunal*), interest is assessed on the correcting amount[21] and paid in the assessment that includes the correcting amount. The interest is assessed from the date when the incorrect amount was certified to the date when the correcting amount is certified.

(3) **ECC2 only**: If the *Project Manager* does not issue a certificate, which the *Project Manager* should issue, interest is paid on the amount that should have been certified.[22] The interest is assessed from the date by which the amount should have been certified until the date when it is certified and is included in the amount then certified. It is possible that this last category of interest might not be enforceable since the *Contractor* might not have suffered any loss. Payment is dependent on the assessment date and not on the certificate date and so a late certificate might not affect payment. In any case, if payment is affected, interest is assessed on the late payment anyway.

or

ECC3 only: If a payment is late because the *Project Manager* did not issue a certificate which he should issue, interest is paid on the late payment.[23] The interest is assessed from the date when the payment should have been made to the date when it was made. This clause is different from the ECC2 clause because it links the lateness of the certificate issue to the lateness of the payment, whereas ECC2 refers only to the lateness of the certificate.

In ECC2, interest is calculated at the *interest rate* stated in Contract Data part one and is compounded annually.[24] In ECC3, interest is calculated on a daily basis at the *interest rate* and is compounded annually.[25]

1.2.3.4 Time period for assessment

The *Project Manager* has less than a week in which to complete his assessment.[26] This short time period may become too onerous for the *Project Manager*, particularly for contracts under Options C, D and E, where there would be insufficient time to examine fully the *Contractor*'s application and all the supporting documentation. It is suggested that in practice the *Project Manager* conducts a spot check of the application for payment and conducts later audits of the *Contractor*'s records which he is required to keep according to the Works Information. This audit process should be detailed in the Works Information and adhered to by the *Project Manager* and his assistants. The *Project Manager* has

---

[19] ECC3 adds the following words to clause 51.2: 'or if a payment is late because the *Project Manager* does not issue a certificate which he should issue'.
[20] Clause 51.2.
[21] Clause 51.3.
[22] ECC2 clause 51.4.
[23] ECC3 clause 51.2.
[24] ECC2 clause 51.5.
[25] ECC3 clause 51.4.
[26] Since he has to issue a payment certificate within one week of each assessment date (clause 51.1).

the ability to correct any wrongly assessed amount due in a later payment certificate,[27] which gives the *Project Manager* the opportunity to make any changes post audit.

1.2.3.5 Details of assessment   The *Project Manager* gives to the *Contractor* details of how the amount was assessed.[28] This is very important as it gives the *Contractor* the opportunity to ascertain how the amount due was assessed if different from his own assessment. This detail would generally be given in or with the payment certificate.

1.2.4 Certification   The *Project Manager* certifies a payment within one week of each assessment date.[29] The *Project Manager* would normally include with the payment certificate details of how the amount due was assessed by him. If any previous amount due was incorrect it is corrected in a later payment certificate. Interest (on previous late payments and on incorrect amounts[30]) is included in later payment certificates.[31] The *Project Manager* should voluntarily include interest in a payment certificate if previous payments were made late to the *Contractor*; the *Project Manager* should not wait for the *Contractor* to claim interest.[32]

1.2.5 Invoices   Any invoice procedure dictated by the *Employer* takes place at this stage.

None of the NEC documents caters for invoicing as part of the payment procedure.

If the *Employer* wishes to receive an invoice from the *Contractor* for the purposes of payment, then an invoice process will have to be introduced into the contract through either secondary Option Z or the Works Information.

Many employers augment the payment procedure in the *conditions of contract* with other information that they prefer to receive and as described in the Works Information. For example, some *Employers* may dictate how the *Contractor* should format his application for payment and what documents should accompany the application such as, in the case of Options C, D or E, labour records, timesheets, plant records, payroll and so on. As part of the payment administration procedures, *Employers* could also include a section on invoicing, instructing the *Contractor* when to submit an invoice and what the invoice should contain. Some examples of what an *Employer* may include in the Works Information regarding invoicing procedures are listed below.

(1)   To whom the invoice is to be addressed. Possibly the *Employer*'s finance department within the *Employer*'s organisation.

(2)   When the invoice should be submitted. Example: Invoices shall be submitted within seven days of the date of the *Project Manager*'s payment certificate or on the date stated on the attached schedule of project dates.

(3)   The sanction on a late invoice. Example: Payment shall be delayed by the number of days that the invoice is late.

(4)   What the invoice should show. Example: Invoices shall show the full amount claimed to date, deducting separately previous payments. The latest statement of account should accompany the invoice. Accounts shall be shown net of VAT, with the amount of VAT shown separately.

Invoicing is usually an internal procedure that facilitates payment and is not therefore strictly speaking a condition of contract. An employer's invoicing procedure normally coincides with the procedures required by their accounts department and may therefore vary from employer to employer. Whatever the invoicing procedure, it takes place within the parameters of the interval between the assessment date and the payment date (or, in the case of Option Y(UK)2 the certificate date and the payment date) and care should be taken to

---

[27] Clause 50.5.
[28] Clause 50.4.
[29] Clause 51.1.
[30] And on late certificates in ECC2 (clause 51.4).
[31] Clause 51.2 and clause 51.3.
[32] An example of a payment certificate is included in Chapter 1 of Book 5.

ensure that a late invoice resulting in a late payment is not construed as an *Employer*'s default.

> The ECC does not include an invoicing procedure.

1.2.6 Payment The *Employer* makes payment a specified time after the assessment date. This may be three weeks,[33] which is the default in the *conditions of contract*, or it may be four or five weeks as included in the Contract Data (see Fig. 1.1).

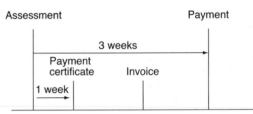

Fig. 1.1. Default time period in the contract

Note that the payment time is dependent on the assessment date and not the date of the payment certificate or the date of the invoice. None of the activities that take place between the assessment date and the payment date should delay payment. Additional clauses using Option Z or additional requirements in the Works Information can be used to ensure that the *Employer* is not in default when a payment is late as a result of a late invoice.

1.2.7 Representation of the payment procedure in the ECC Figure 1.2 assumes that certified payment is made within four weeks of the assessment date, rather than the default three weeks in clause 51.2 of the ECC. An *Employer* would need to include the time period (four weeks) in Contract Data part one to effect this change.

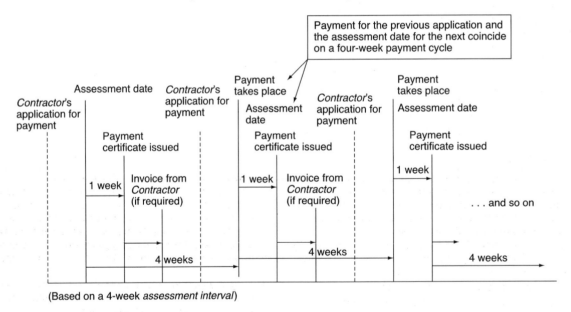

Fig. 1.2. Payment procedures in the ECC

1.3 How the ECC3 includes the HGCR Act 1996

ECC3 incorporates the Housing Grants, Construction and Regeneration (HGCR) Act 1996 in two different places, whereas ECC2 just uses Option Y(UK)2. ECC3 incorporates payment changes in Option Y(UK)2 and adjudication changes in Option W2.

---

[33]Clause 51.2.

ECC3 incorporates and approaches the requirements of the HGCR Act 1996 in a far simpler and more easily comprehensible manner, whereas in ECC2 there was an addendum which amended, added, deleted or replaced core clauses. ECC3 has a redrafted option Y(UK)2 which sets out in the clauses Y2.1, Y2.2 and Y2.3 the requirements of the Act.[34]

Table 1.1 shows how ECC3 option Y(UK)2 is set out and how it fulfils the requirements of the HGCR Act 1996.

Table 1.1 ECC3 and Option Y(UK)2

Housing Grants, Construction and Regeneration Act 1996[a]	Option Y(UK)2: the Housing Grants, Construction and Regeneration Act 1996
Section 116 Reckoning periods of time (3) Where the period would include Christmas Day, Good Friday or a day which under the Banking and Financial Dealings Act 1971 is a bank holiday in England and Wales or, as the case may be, in Scotland, that day is excluded.	**Y2.1(2) Definitions** 'A period of time stated in days is a period calculated in accordance **with Section 116 of the Act**'

Section 110 Dates for payment
(1) Every construction contract shall
   (a) provide an adequate mechanism for determining **what payments become due under the contract, and when**, and
   (b) provides for a **final date for payment** in relation to any sum which becomes due.
The parties are free to agree how long the period is to be between the date on which a sum becomes due and the final date for payment.
(2) Every construction contract shall provide for the giving of notice by a party not later than five days after the date on which payment becomes due from him under the contract, or would have become due if
   (a) the other party had carried out his obligations under the contract, and
   (b) no set-off or abatement was permitted by reference to any sum claimed to be due under one or more other contracts
specifying the amount (if any) of payment made or proposed to be made, and the basis on which that amount was calculated.
(3) If or to the extent that a contract does not contain such provision as is mentioned in subsection (1) or (2), the relevant provisions of the Scheme for Construction Contracts apply.
	Y2.2 Dates for Payment
	'**The date on which a payment becomes due is seven days after the assessment date**' (the latest date for payment under clause 51.1).
	
	'The **final date for payment is fourteen days or a different period for payment if stated in Contract Data** after the date on which payment becomes due' (the latest date for payment under clause 51.2).
	
	ECC3 Payment Certificate (clause 51.1 and Y2.2)

Section 111 Notice of intention to withhold payment
(1) A party to a construction contract **may not withhold payment after the final date for payment of a sum due under the contract unless he has given an effective notice of intention to withhold payment.**
The notice mentioned in section 110(2) may suffice as a notice of intention to withhold payment if it complies with the requirements of this section.
(2) To be effective such notice **must specify**
   (a) **the amount proposed to be withheld and the ground for withholding payment**, or
   (b) **if there is more than one ground, each ground and the amount attributable to it**,
and must be given not later than the prescribed period before the final date for payment.
(3) The parties are free to agree what the prescribed period is to be.
In the absence of such agreement, the period shall be that provided by the Scheme for Construction Contracts.
	Y2.3 Notice of intention to withhold payment
	'If either Party intends to withhold payment of an amount due under this contract, he notifies the other Party not later than seven days (**the prescribed period**) before the final date for payment **by stating the amount proposed to be withheld and the reason for withholding payment. If there is more than one reason, the amount for each reason is stated.**
	
	A Party does not withhold payment of an amount due under this contract unless he has notified his intention to withhold **payment** as required by this contract.'

[34] For a more detailed review of the HGCR Act 1996 see Chapter 4.

Table 1.1 (*Continued*)

Housing Grants, Construction and Regeneration Act 1996[a]	Option Y(UK)2: the Housing Grants, Construction and Regeneration Act 1996
Section 112 Right to suspend performance for non-payment (1) Where a sum due under a construction contract is not paid in full by the final date for payment and no effective notice to withhold payment has been given, the person to whom the sum is due **has the right (without prejudice to any other right or remedy) to suspend performance of his obligations under the contract** to the party by whom payment ought to have been made ('the party in default'). (2) The right may not be exercised without first giving to the party in default at least seven days' notice of intention to suspend performance, stating the ground or grounds on which it is intended to suspend performance. (3) The right to suspend performance ceases when the party in default makes payment in full of the amount due. (4) Any period during which performance is suspended in pursuance of the right conferred by this section shall be disregarded in computing for the purposes of any contractual time limit the time taken, by the party exercising the right or by a third party, to complete any work directly or indirectly affected by the exercise of the right. Where the contractual time limit is set by reference to a date rather than a period, the date shall be adjusted accordingly.	**Y2.4 Suspension of performance** 'If the *Contractor* exercises his **right under the Act to suspend performance** it is a compensation event.'

[a]Extract from Housing Grants, Construction and Regeneration Act 1996.

1.4 The effects of Y(UK)2[35]

1.4.1 Periods for payment   Where the *Employer* has determined that the HGCR Act 1996 applies to the contract, and thus has chosen Option Y(UK)2 to apply to the contract, the payment procedure is affected.

Two new concepts are introduced:

(1)   the date on which a payment becomes due and
(2)   the final date for payment.

The date on which the payment becomes due is not the date on which payment has to be made but rather the date which marks that the payment is due to be paid some time in the future. The date on which a payment becomes due is seven days after the assessment date[36] and is the latest day by which the *Project Manager* must certify a payment.[37]

> The due date is different from the final date for payment.

The final date for payment is the latest date by which each certified payment must be made. The final date for payment is a certain time period after the date on which the payment becomes due.[38] The final date for payment is differ-

---

[35]Note that Y(UK)2 in ECC2 is very different from Y(UK)2 in ECC3, although both deal with payment.
[36]ECC2 clause 56.1 of Y(UK)2; ECC3 clause Y2.2 of Option Y(UK)2.
[37]ECC2 clause 51.1 of Y(UK)2; ECC3 clause Y2.2 of Option Y(UK)2.
[38]Rather than a certain period after the assessment date, as with the default *conditions of contract*.

ent in ECC2 and ECC3. In ECC2, the time period is either 21 days as default, or a different period if stated in the Contract Data.[39] In ECC3, the final date for payment is 14 days after the date on which payment becomes due, or a different period if stated in the Contract Data.[40]

1.4.1.1 Differences between Y(UK)2 and the ECC section 5

The certification procedure is unchanged since both the ECC and Y(UK)2 require the *Project Manager* to issue a certificate within seven days of the assessment date.

(1)   In ECC2, the payment date (final date for payment in Y(UK)2) is longer in Y(UK)2 than in the ECC section 5.

Clause 51.2 requires payment to be made within three weeks after the assessment date. The three weeks time period is a default that can be changed using an optional statement in the Contract Data. The important point is that the trigger is the assessment date (see Fig. 1.3).

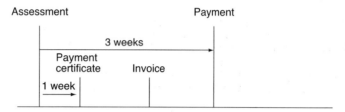

Fig. 1.3. Times for payment in the ECC section 5

Clause 56.1 (clause Y2.3 of Y(UK)2) requires payment to be made on the final date for payment, which is three weeks after the date on which payment becomes due (which is seven days after the assessment date). Once again, the three weeks time period is a default that can be changed using an optional statement in the Contract Data. This time, however, the trigger is when payment becomes due – that is, certification. (See Fig. 1.4.)

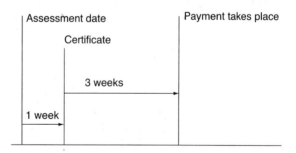

Fig. 1.4. Times for payment in Y(UK)2

In ECC3, there is no difference between the payment date and the final date for payment in Option Y(UK)2. This is because the final date for payment is 14 days after the date on which payment becomes due, whereas in clause 51.2, payment is made three weeks after the assessment date. Because the date on which payment becomes due is seven days after the assessment date, the time periods between assessment and payment are the same, although the time periods run from different dates. See Fig. 1.5.

---

[39] ECC2 clause 56.1 of Y(UK)2.
[40] ECC3 clause Y2.2 of Option Y(UK)2.

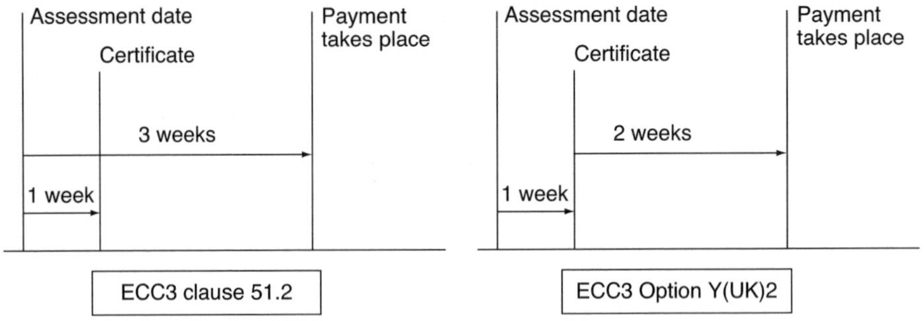

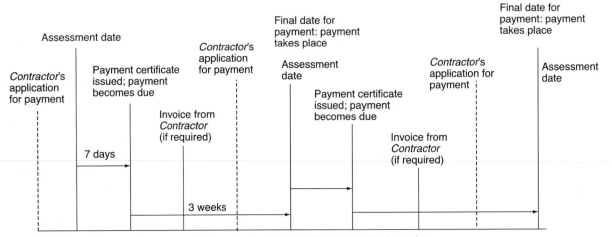

Fig. 1.5. Times for payment in ECC3 clause 51.2 and Option Y(UK)2

1.4.2 Payment procedure with Y(UK)2 — The payment procedure with Y(UK)2 is shown schematically in Fig. 1.6.

Fig. 1.6. Payment procedure with Y(UK)2

(Based on a four-week *assessment interval* and a final date for payment 21 days after the date on which payment becomes due)

1.4.3 Withholding payment with Y(UK)2 — Option Y(UK)2 also introduces the concept of withholding payment to comply with the Act. If the *Employer* intends to withhold payment from the *Contractor*, then the *Contractor* should be informed about the amount that is to be withheld and the reason for withholding this amount. The *Employer* must tell the *Contractor* this not later than seven days before the final date for payment.[41]

The *Contractor* may suspend performance if the *Employer* does not issue a notice of withholding and payment is not made in full by the final date for payment (see Fig. 1.7).

1.5 Payment procedure on Completion

An assessment date occurs at Completion of the whole of the *works*.[42] The rest of the procedure is the same as with interim payments.

It is possible that a *Contractor* might inform the *Project Manager* that Completion has been achieved, although this is not part of the procedure of the ECC. The *Project Manager* is at liberty to ignore such a communication or to acknowledge it, although in the interests of mutual trust and cooperation, an answer would be advised. It is the *Project Manager* who decides that Completion has

---

[41]ECC2 clause 56.2 of Y(UK)2; ECC3 clause Y2.3.
[42]Clause 50.1.

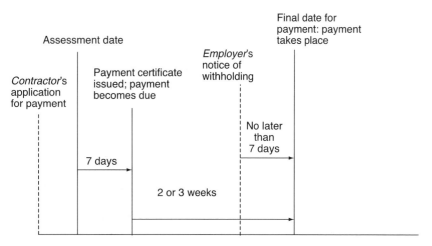

Fig. 1.7. Withholding payment procedure

been achieved, based on the objective criteria stated in the Works Information and once the *Project Manager* has notified the *Contractor* that Completion has taken place, an assessment can take place. This assessment could be slightly delayed if the *Project Manager* first requests an application for payment on which to base the assessment, and this can take place by agreement between the *Project Manager* and the *Contractor* based on common sense.

An assessment at Completion would take into account, in particular, any retention to be released if the relevant secondary Option[43] has been chosen as part of the contract.

1.6 Payment procedure after Completion

In ECC2, an assessment date takes place after Completion of the whole of the *works* only if an amount due is corrected or if a payment is made late.[44] In other words, if the amount on a previous payment certificate is corrected after Completion, perhaps by the *Project Manager* as a result of a mistake or following a decision of the *Adjudicator* or *tribunal*,[45] then an assessment takes place in order to correct the amount due and to cater for interest on the incorrect amount.

If a previous payment was late then the *Project Manager* should instigate a further assessment in order to pay interest to the *Contractor* on the amount that was late.[46]

In ECC3, assessments take place right through Completion and the period up to the *defects date*.[47] As a result, there is no need to have a special assessment to correct amounts due since regular assessments would be taking place as a matter of course.

1.7 Payment procedure after the *defects date*

An assessment date takes place four weeks after the *Supervisor* issues the Defects Certificate.[48]

The Defects Certificate is either a list of Defects that the *Supervisor* has notified before the *defects date* which the *Contractor* has not corrected or, if there are no such Defects, a statement that there are none.[49] The *Supervisor* issues the

---

[43] ECC2 Option P; ECC3 Option X16.
[44] Clause 50.1.
[45] Clause 51.3.
[46] Clause 51.2.
[47] ECC3 clause 50.1.
[48] Clause 50.1.
[49] ECC2 clause 11.2(16); ECC3 clause 11.2(6).

Defects Certificate at the later of the *defects date* and the end of the last *defect correction period*.[50]

This assessment would be related to the release of retention and the deduction of monies where the *Contractor* has not corrected Defects that were required to be corrected.

---

[50]ECC2 clause 43.2; ECC3 clause 43.3.

---

2 Control of Time

Synopsis

This chapter discusses aspects relating to the *Contractor*'s programme including:

- The terminology used to describe the programme

- What the programme is

- The definition and purpose of the Accepted Programme

- How and when to submit programmes

- What the programme is used for

- What to include in the programme

2.1 Introduction

One of the underlying principles of the ECC is to avoid and reduce the amount of change that occurs on construction projects. However, the ECC contract recognises that change is inevitable even when the project has been well planned and prepared. Having accepted that change is inevitable, the contract sets out to deal with the effects and consequences of change in an improved way.

The programme is one of the most important tools for use by both the *Contractor* and the *Project Manager* throughout the duration of the project. It is valuable not only as a scheduling tool but also as a pricing and project management tool and a change control tool. Indeed, there is an adage, which states that 'planning plus management equals project management'.

The ECC has unique features for dealing with the programme and is a manual of procedure for the Parties to the contract. The ECC programme is the *Contractor*'s programme. He compiles the programme and he owns the terminal float, but the programme is also a project management tool, and there are incentives within the ECC to ensure that the *Contractor* keeps his programme up to date. The programme is used not only as a project planning tool, but can also be used for forensic analysis to assess the effects of a compensation event on time and money.

Although the clauses for programming are to be found primarily within section 3 of the ECC core clauses, other clauses might impact or be affected by the programme. The ECC should be read as a whole to be properly understood and implemented.

> The programme is an important tool for use by the *Contractor* and *Project Manager* to manage the contract.

2.2 Terminology

The following words or phrases denote the terminology in the ECC that affects the programme.

- Contract Date   defined in ECC2 clause 11.2(3); (ECC3 11.2(4))
- *starting date*   identified in Contract Data part one
- *possession date*[51]   identified in Contract Data part one
- Completion   defined in clause 11.2(13) (ECC3 11.2(2))
- Completion Date   defined in clause 11.2(12) (ECC3 11.2(3))
- planned Completion   used in clause 31.2 and 63.3
- Key Date   ECC3 specific and defined in clause 11.2(9).

2.2.1 Contract date   This is the date when the contract came into existence and is generally stated in the Articles of Agreement/Form of Contract/Letter of Award.

2.2.2 *Starting date*   The date stated in the Contract Data part one, when the *Contractor* starts work on the contract.

In effect the *starting date* is the date from which the *Contractor* is 'on risk'.[52] The *starting date* is not necessarily the date when the *Contractor* will commence work on Site, since the work commenced at the *starting date* could be design or manufacture required to be undertaken off-site prior to construction or enabling works required on Site.

2.2.3 ECC2 *possession date(s)* – ECC3 *access date(s)*   The date or dates stated in the Contract Data part one, before which the *Contractor* cannot start work **on the Site**.

The *starting date* is not necessarily the same as the *possession date*[53] because the *starting date* is not necessarily the date when the *Contractor* will take

---

[51]Called *access date* in ECC3.
[52]Clause 81.1.
[53]Called *access dates* in ECC3.

possession[54] of the site. This is because, on contracts which have a significant amount of design or manufacturing work for the *Contractor*, the *starting date* could be several weeks or months prior to starting work on Site to enable the *Contractor* to design or manufacture the Plant and Materials which are required to Provide the Works.

The *Contractor* might not have sole possession/access of the Site. If so, this should be stated in the Works Information as well as any interfaces required.

> ECC3 uses the words '*access date*' in lieu of '*possession date*'.

2.2.4 Completion   Completion is when the *Contractor* has done all the work which the Works Information states he is to do by the Completion Date and has corrected Defects which would have prevented the *Employer* from using the *works*.[55] ECC3 has extended the coverage of this clause by adding the words, 'or Others[56] from doing their work'. This recognises that the *Employer* may employ others to undertake works in association with the project and that they may be delayed by uncorrected Defects.

Completion is not a date but a status to be achieved.

> Completion is not a date.
>
> Completion could occur on the Completion Date, after the Completion Date or before the Completion Date.
>
> The date of Completion is decided by the *Project Manager*.

2.2.5 Completion date   The Completion Date is the date stated in the Contract Data as the *completion date* unless later changed in accordance with the contract; that is, it is the date on or before which the *Contractor* is contractually obliged to complete the *works* if he is not to be in default.

The Completion Date could be different from the date of Completion, where the date of Completion is that date that the *Project Manager* decides the *works* are complete.

2.2.6 A Key Date (ECC3-specific)[57]   A Key Date is the date by which work is to meet the Condition stated in the Contract Data unless the Key Date or Condition is later changed in accordance with the contract.[58] There is a sanction on the *Contractor* if the work does not meet the Condition stated for a Key Date and the *Employer* incurs additional cost as a result.[59]

2.2.7 Secondary Option L[60] – sectional completion   This Option needs to be chosen where the *Employer* wishes to take over any part of the *works* before Completion of the whole of the *works*.

2.2.8 Planned Completion   Planned Completion is the date when the *Contractor* plans to complete the *works*. He is required to show both planned Completion and the Completion Date on his programme.

At the outset of the contract the *Contractor*'s planned Completion will be a date earlier than the Completion Date given by the *Employer* in Contract Data part one. During the contract the *Contractor*'s planned Completion and the Completion date will be adjusted. The Completion date will be adjusted for admissible

---

[54]Called access in ECC3.
[55]ECC2 clause 11.2(13); ECC3 clause 11.2(3).
[56]ECC2 clause 11.2(2) and ECC3 clause 11.2(10), Others is a defined term meaning people and organisations who are not parties to the contract.
[57]ECC3 clause 11.2(9).
[58]ECC3 clause 11.2(9).
[59]ECC3 clause 25.3.
[60]ECC3 – option X5.

compensation events to give a revised Completion date and *Contractor's* planned Completion. The *Contractor's* planned Completion will also be adjusted for matters which are at the *Contractor's* risk under the contract. It is therefore possible for the planned Completion to be later than the Completion Date, which may expose the *Contractor* to delay damages.

2.3 What is the programme?

The programme is not a key feature in JCT or ICE Contracts. Traditionally, an air of mistrust seems to surround the preparation and review of programmes with the *Contractor* not wishing to show too much information on the programme for fear that the Engineer or Architect will manipulate it to his advantage in the event that variations or other changed circumstances arise for additional or reduced works.

With traditional contracts the contractor is not particularly motivated to prepare and regularly update programmes for joint agreement with the employer's representative since there are little or no sanctions that can be applied to the contractor apart from warnings that the lack of an approved programme may jeopardise the assessment of claims for extensions of time.

The fact that the programme may be submitted at tender stage and is not updated or referred to again in traditional contracts denies the project manager access to a valuable tool.

> The programme is regularly updated, unlike traditional forms of contract.

The programme in the ECC is far more than a simple bar chart showing the intended order and duration of activities to be completed. The programme includes the resources that are used for each activity, including labour and Equipment, and therefore becomes a base comparison for any compensation events. It is a project management tool and as such is updated regularly.

An example of a traditional clause 14 programme from an ICE 6th Edition contract is included in Fig. 2.1. Compare the lack of information on this programme to the requirements of the ECC.

2.4 Definition of the Accepted Programme

The Accepted Programme is the latest programme accepted by the *Project Manager*.[61]

> The *Project Manager* decides whether to accept a programme submitted by the *Contractor*.

The programme is submitted by the *Contractor* for acceptance. Upon its acceptance, it becomes the Accepted Programme. Each subsequent programme submitted by the *Contractor* to the *Project Manager* becomes the Accepted Programme upon acceptance, superseding the previous programme.

2.4.1 When the programme is submitted for acceptance

The *Employer* may choose to receive a first programme from the *Contractor* either at tender stage or after contract award.

(1)   Usually the *Employer* will request a programme to be submitted by the *Contractor* with his tender in the detail required in clause 31.2. In this instance the *Contractor* will refer to the programme in part two of the Contract Data which he completes at the time of tender.

(2)   The alternative is to let the *Contractor* submit a programme within a period identified by the *Employer* in Contract Data part one, this period

---

[61]ECC2 clause 11.2(14); ECC3 clause 11.2(1).

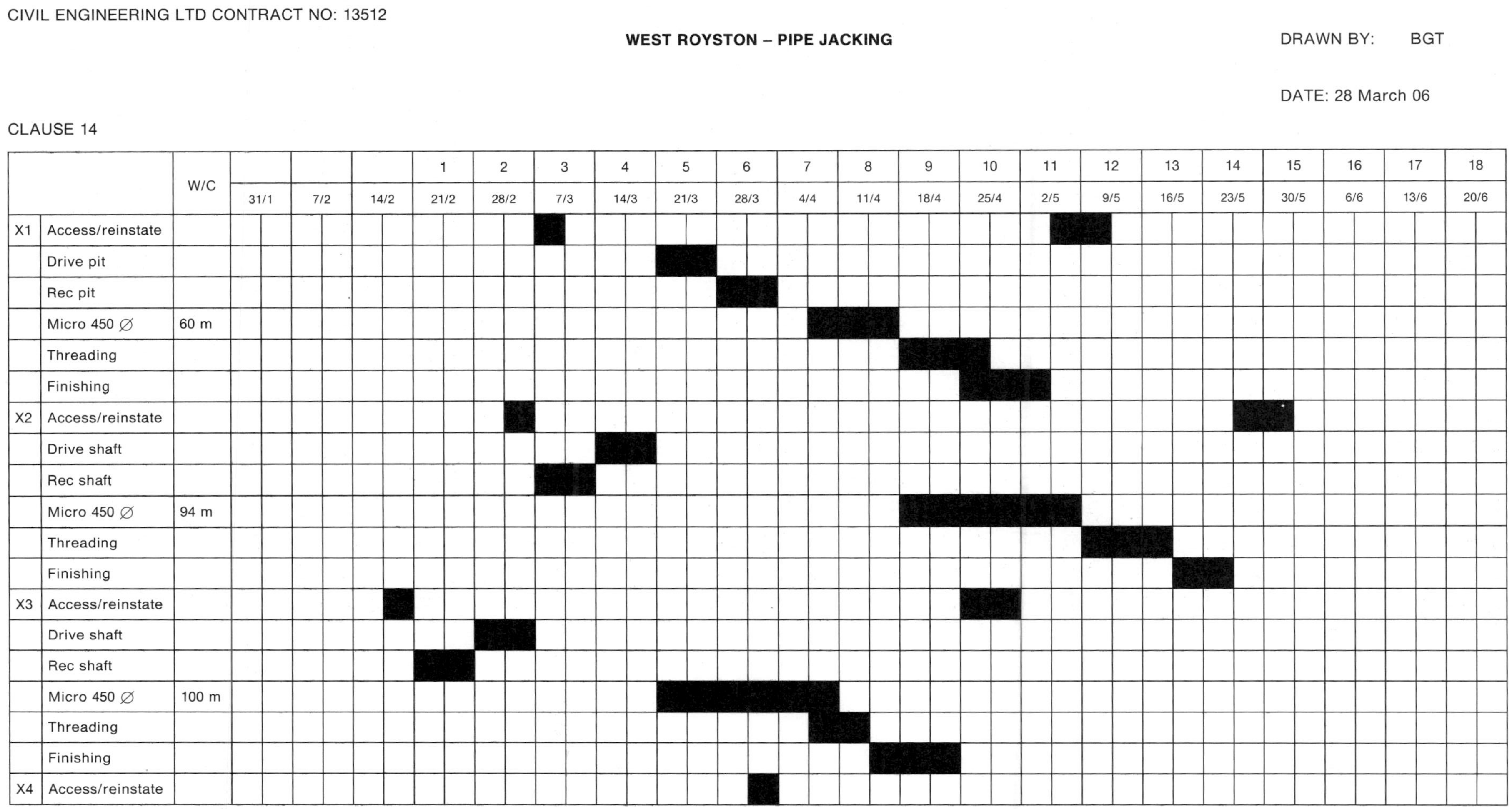

CIVIL ENGINEERING LTD CONTRACT NO: 13512

WEST ROYSTON – PIPE JACKING

DRAWN BY: BGT

DATE: 28 March 06

CLAUSE 14

	W/C	31/1	7/2	14/2	21/2	28/2	7/3	14/3	21/3	28/3	4/4	11/4	18/4	25/4	2/5	9/5	16/5	23/5	30/5	6/6	13/6	20/6
							1	2	3	4	5	6	7	8	9	10	11	12	13	14	15	16	17	18
X1	Access/reinstate																					
	Drive pit																					
	Rec pit																					
60 m	Micro 450 ⌀																					
	Threading																					
	Finishing																					
X2	Access/reinstate																					
	Drive shaft																					
	Rec shaft																					
94 m	Micro 450 ⌀																					
	Threading																					
	Finishing																					
X3	Access/reinstate																					
	Drive shaft																					
	Rec shaft																					
100 m	Micro 450 ⌀																					
	Threading																					
	Finishing																					
X4	Access/reinstate																					

Fig. 2.1. Traditional clause 14 programme

commencing at the Contract Date. This alternative is included because it is recognised in the ECC that in certain instances, for example on a cost-reimbursable contract, a fully developed and detailed programme may not be possible prior to award of the contract. This alternative therefore provides flexibility.

Where possible, however, it is a good policy to have a programme submitted and agreed before contract award as this puts both Parties in a far better position in terms of what is required and when. It also enables the *Employer* to ensure that the information and dates stipulated in the tender have been adhered to. This is particularly relevant for priced contracts. With cost-reimbursable contracts the programme is likely to develop over the course of the project and therefore it may not always be possible for a detailed programme to be submitted at tender stage.

> The *Employer* could request a first programme from the *Contractor* either at tender stage or after contract award.

2.5 The purpose of the Accepted Programme

In keeping with good project management practice, the ECC recognises the programme as being an essential tool for managing the *works*. It enables the *Project Manager* and the *Contractor* to monitor progress and to assess the delaying effects of any compensation events that arise. It also enables the *Project Manager* to see what the time effects will be of a proposed instruction that he might be considering.

It is important always to know what the Completion Date is, if the *Employer*'s right to delay damages is not to be frustrated. Good programme management by the *Contractor* could protect him from delay damages being levied by the *Employer*.

The programme is used for more than just tracking progress of the project, although this is clearly its primary function.

2.5.1 What the Accepted Programme can be used for

The programme can be used for the following.

2.5.1.1 Resources

The programme includes a resource statement;[62] that is, a list for each activity of the resources that are intended to be used. Clearly this list will initially be based on the scope of work at the time that the programme is drafted – either at tender stage, or shortly after the start of the contract. This resource statement then becomes useful during the project.

The resources facilitate the estimating of the job by the *Contractor*. Since he will be going through the exercise of forecasting the resources used for the job, this is smoothly translated into the programme.

> The programme is an essential part of assessing compensation events.

The resources facilitate the assessment of compensation events, both for the *Contractor* and the *Project Manager*. The resources and duration that were expected for the activity can be used as a base when considering the changes as a result of a compensation event. The *Contractor* is able to use those resources estimated in considering the impact of the compensation event, and the *Project Manager* can assess the quotation in the light of the resources estimated originally.

---

[62]Usually included with the method statement. In ECC3 this requirement has been amended to, 'statement of how the *Contractor* plans to do the work'.

2.5.1.2 Costs   Because of the resources attached to the programme, the costs of the project become clearer and the *Project Manager* is able to view the programme both from a scheduling point of view and a cost perspective.

This becomes particularly important where the *Project Manager* assesses a compensation event,[63] since he will use the tools available to him to do so, and may therefore use the latest Accepted Programme to assess the changes to the Prices as well as the delay to the Completion Date.

2.5.1.3 Project management   The progress of the project is noted on each revised programme so that the *Project Manager* is able to project manage the project through the regularly updated programmes by noting whether the project is progressing on time and to budget, using not only the scheduling aspect of the programme but also the resource and cost aspects.

The effect of a compensation event on the programme is more easily assessed since the *Contractor* can immediately see the impact of any delays on his programme. If the *Project Manager* assesses the programme,[64] he will use those tools available to him, including the *Contractor*'s programme. If the programme is not up to date, the *Project Manager* could assess the compensation event based entirely on his own experience. The *Contractor* is thus incentivised to keep his programme up to date and accurate.

2.6 What is included in the programme?

2.6.1 All programmes   Each programme that is submitted for acceptance must include the following information.[65]

(1)   Dates:
  • *starting date*,
  • *possession dates*,[66]
  • Key Dates – (ECC3 only),
  • planned Completion,
  • Completion Date,
  • dates when the *Contractor* plans to meet each Condition stated for the Key Dates (ECC3 only),
  • dates when the *Contractor* plans to complete work needed to allow the *Employer* and Others to do their work,
  • dates when the *Contractor* will need:
    • possession (if later than the *possession date*),[67]
    • acceptances,
    • Plant and Materials and other things to be provided by the *Employer*,
    • Information from others (ECC3 only).

(2)   The order and timing of the operations the *Contractor* plans to do. Operations may be things which the Contractor has to undertake in order to do the *work*. For example, the design or manufacture of bathroom pods for hotels which will be manufactured/designed off-Site, with the *work* on Site being their positioning in place by crane and final connections and testing.

(3)   The order and timing of the work of the *Employer* and Others as included in the Works Information or as later agreed with them. In ECC3 the wording has been amended to 'the order and timing of the work of the *Employer* and Others as last agreed with them by the *Contractor* or, if not so agreed, as stated in the Works Information.'

(4)   A method statement for each operation. ECC3 deletes the words 'method statement' and replaces them with 'a statement of how the *Contractor* plans to do the work'. See the following item.

[63] Clause 64.
[64] Under clause 64.
[65] Clause 31.2.
[66] Called *access dates* in ECC3.
[67] ECC3 – 'access to a part of the Site if later than its *access date*'.

(5)   A resource statement for each operation. In ECC3 this requirement has been amended as follows: 'for each operation, a statement of how the *Contractor* plans to do the work identifying the principal Equipment and other resources which he plans to use'.

(6)   Provisions for:
  • float,
  • time risk allowances,
  • health and safety requirements,
  • the procedures set out in the contract.

(7)   Other information that the Works Information requires the *Contractor* to show on a programme submitted for acceptance.

2.6.2 All programmes except the first programme

In addition to the above information, each revised programme must also include the following information:[68]

  • the actual progress achieved on each operation and its effect upon the timing of the remaining work,
  • the effects of implemented compensation events and of notified early warning matters,
  • how the *Contractor* plans to deal with any delays and to correct notified Defects,
  • any other changes which the *Contractor* proposes to make to the Accepted Programme.

The first bullet point is essential for facilitating the *Project Manager*'s reviewing of progress.

The final bullet point is of interest because it allows the *Contractor* to reprogramme *works* to suit any changes he might have with regards to how he will Provide the Works.

A *Contractor* could find himself in a position where he realises that his original planned sequence of operations is no longer realistic or practicable and he decides to resequence the work. If he does this, and he is at liberty to do so, under Options A and C the revised list of new or amended activities and the programme must correlate. The prices of the individual activities will also need to be amended so that they tie up.

2.6.3 Notes on the programme inclusions

2.6.3.1 Starting date, possession/ access dates, Completion Dates and Key Dates (ECC3 only)

The *starting date, possession/access dates* and Completion Date are all stated in Contract Data part one and should be included in the programme. ECC3 also requires Key Dates to be shown as well as the Condition to be met by that Key Date.

2.6.3.2 Planned Completion

Planned Completion is required to be shown on the programme separately from the Completion Date. Planned Completion shows that date when the *Contractor* is planning to complete the *works*. The date for planned Completion will at the outset of the contract be a date earlier than the Completion Date.

During the contract however the *Contractor* may experience problems or encounter risks which are at his risk under the contract and this may lead to planned Completion being later than the original or adjusted contract Completion Date. In such instances the *Contractor* will be at risk for delay damages and he will be required to show on revised programmes how he plans to recover the delay.

The terminal float (the period between planned Completion on the programme and the Completion Date) is retained by the *Contractor*, as stated in clause 63.3,[69] where any delay to the Completion Date due to a compensation event is assessed as the length of time that planned Completion is later than planned Completion on the Accepted Programme.

[68] Clause 32.1.
[69] ECC3 – delays to Key Dates assessed on the same basis.

However, even now that there is a facility for planning to complete earlier than the Completion Date stated in the contract, few contractors appear to be including a planned Completion date in their programme.

It should also be pointed out that the Completion Date for the contract is something different from Completion. Completion is a status[70] that is achieved when the *Contractor* has fulfilled his duties as described in the contract. Completion could therefore be achieved on, before, or after the Completion Date.

ECC3 introduces the concept of Key Dates[71] and states: 'A delay to a Key Date is assessed as the length of time that, due to the compensation event, the planned date when the Condition stated for a Key Date will be met is later than shown on the Accepted Programme.'

> In ECC3 the wording of clause A54.2 and C54.2 has been revised to read as follows:
>
> 54.2 'If the *Contractor* changes a planned method of working at his discretion so that the activities on the Activity Schedule do not relate to the operations on the Accepted Programme, he submits a revision of the Activity Schedule to the *Project Manager* for acceptance.'

2.6.3.3 Other dates to be shown on the Programme

It is important for the *Employer* to advise the *Contractor* of the dates when he requires the *Contractor* to complete particular works (but for **use not take-over**, i.e. not a *section* of the *works*) or the degree to which he needs it to be completed to enable him and Others, whose names or allocations must also be included in the documents, to do their work.

> An example may be a statement such as:
>
> The *Contractor* shall complete the proposed new road between chainages 0 to 100 m by the 10th December 2006 to allow access for the *Employer* to the existing warehouse adjacent to the new road. It is a requirement that the road be complete up to base course level with all adjacent kerbs, drainage and street lighting.

The statement needs to be clear and unambiguous: the *Contractor* should not be in any doubt as to what it is he has to do by the key date.[72] On an ECC3 contract this would be an example of the *condition* of work to be completed by a *key date*.

2.6.3.4 Possession/access of a part of a Site if later than its possession/access date

The importance of this is that the Contract Data part one should include the dates when the *Employer* can give possession/access of parts of the Site.

These dates have to be shown on the *Contractor*'s programme. If the *Contractor* decides that he does not need to take possession/access of a part of the Site on the *possession/access dates* given then he can decide to include a later *possession date* into his programme. The *Employer*'s contractual obligation to give possession/access changes to this later date shown on the *Contractor*'s programme.

Figure 2.2 shows in a simple form the consequences of giving a later *possession/access date* than that given in the Contract Data part one.

Line A shows the dates for the *starting date, possession/access date* and Completion Date, which the *Contractor* must show on his programme.

---

[70]For more information, see Chapter 1 of Book 1.

[71]Which are defined in ECC3 clause 11.2(9) as 'the date by which work is to meet the Condition stated'.

[72]ECC3 – the term Key Date(s) is a defined term and the *Employer* identifies the condition of work to be completed by the *key date* in the Contract Data.

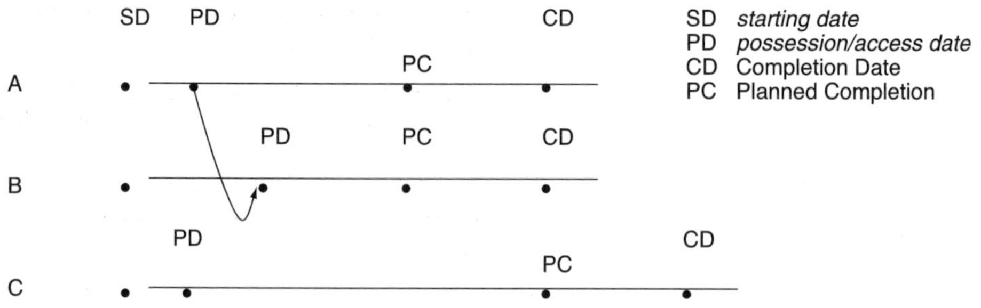

Fig. 2.2. Possession/access dates

Line B shows the *starting date, possession/access date* and Completion Date on the *Contractor*'s programme. He therefore shows the *possession/access date* and his planned Completion for him to be able to Provide the Works.

You can see that the *possession date/access date* in Line B comes later than in Line A and therefore becomes the contractual *possession/access date*.

Line C shows a *possession date/access date* earlier than the *possession/access date* given in the Contract Data and a Completion Date later than that given in the Contract Data. The *Project Manager* will not accept this programme because it does not show the information, which this contract requires, and because it does not allow the *Employer*, other contractors and Others to start, carry out and complete their works as they intended and as stated in the Works Information.

2.6.3.5 Acceptances   If the *Contractor* is designing part or the whole of the *works*, then the *Contractor* will need to show on his programme the dates when he requires his *Contractor* design to be accepted by the *Project Manager*, which includes both permanent and temporary works.

In some instances the Works Information may contain a requirement for the *Contractor* to submit particulars of the design of an item of Equipment (e.g. temporary shoring to a building façade, crash deck). The *Employer* may feel that it is wise to do this because the failure of such Equipment may potentially pose a significant health, security, safety, operational, environmental, public or people fatality risk.

In such instances the *Employer* will wish to demonstrate that if such failure occurs that he acted in a manner which mitigated or reduced such risks. The *Project Manager* may also (clause 23.1) instruct the *Contractor* to submit particulars of his design during the contract. Again in such instance this will need to be shown on any revised programmes submitted by the *Contractor* for acceptance by the *Project Manager*.

On some projects the acceptances given by the *Employer* will need to be approved by third parties – for example, the utilities or statutory bodies. In these instances the *Employer* needs to ensure that he has adequate time allowances built into the acceptance system for *Contractor*-designed work. It also needs to be clear in the Works Information who has the responsibility to liaise and interface with such third parties.

2.6.3.6 Plant and Materials and   The *Contractor* must show on his programme the dates when he requires Plant
other things   and Materials which the Works Information states the *Employer* is to provide.

This will cover situations where the *Employer* might be a company that has items such as standard fittings or furnishings that they have for all their stores, factories or facilities. In some instances the Plant and Materials will be available or will be provided from the *Employer*'s own stores.

A practical example of this is where the *Employer* has placed a contract with a *Contractor* to carry out some utility diversion work. In order for the overall programme for the *works* to be achieved, it was necessary for the *Employer* to pre-order the pipework and fittings because they are on a long lead-in time to purchase of 18 weeks.

The Works Information should also state a date after which the *Employer*-supplied materials will be available for call-off by the *Contractor*.

2.6.3.7 Order and timing of the Contractor's work

The 'operations' referred to in clause 31.2 do not necessarily equate to an *activity schedule* or *bill of quantities* item as there may be several operations to such an item.

For example, there may be an item in the *activity schedule,* which reads:

Activity No.	Description	Price
300	Foundations up to DPC Level	£25,000

This one activity has several operations:

(1)   Excavate for strip foundations.
(2)   Earthwork support.
(3)   Dispose of excavated material.
(4)   Level and compact bottom of excavations.
(5)   Concrete strip foundations.
(6)   Brickwork, including dpc, cavity fill.
(7)   Backfilling, including level and compact.

Compatibility of the *activity schedule* and the Accepted Programme

It is essential that the *activity schedule* and the Accepted Programme or the first programme submitted for acceptance correlate and that each has the same list of activities.

In ECC2, the *Contractor* must show the start and finish of each activity on the *activity schedule* on each programme that he submits for acceptance. In ECC3, the *Contractor* must provide information which shows how each activity on the *activity schedule* relates to the operations on each programme which the *Contractor* submits for acceptance.

If the documents received from the *Contractor* are not compatible then they should be made to be so. Subsequently, clause A54.2/C54.2 of the contract requires that where a *Contractor* changes a planned method of working at his discretion so that the *activity schedule* does not comply with the Accepted Programme, he submits a revision of the *activity schedule* to the *Project Manager* for acceptance.

Incompatibility is a reason for the *Project Manager* not accepting the programme.

If the documents do not correlate then it will be impossible to monitor the true effects that compensation events will have (or have had) upon the Accepted Programme and hence the time aspects of the project will not be manageable.

2.6.3.8 The order and timing of the Employer's part of the works

The *Employer* needs to ensure that any constraints on how the *Contractor* Provides the Works are stated in the Works Information. The *Contractor* will then need to reflect these constraints in his planning of the order and timing of the *works*.

To introduce constraints at a later date once work has commenced would be a change to the Works Information and consequently a compensation event.

The same rule applies for:

• any work which the *Employer* is to carry out,
• any work to be undertaken by Others,
• key dates[73] by which the *Employer* or Others need to complete their work.

This emphasises the requirement for the *Employer* to provide precise and accurate information.

[73]ECC3 – Key Date is a defined term see clause 11.2(9).

2.6.3.9 The importance of dates in the programme

The *Contractor*'s programme is required to show not only the order and timing of his own work, but also that of the *Employer* and Others, and the dates when the *Contractor* needs possession/access, acceptances and things to be provided by the *Employer*. This allows the *Contractor* to plan his operations and to be proactive in his requirements. It enables the *Contractor* to assess any impact on the programme of the *Employer* not doing his work, or not providing something for any resulting compensation event.

The dates to be included in the programme (such as dates when the *Contractor* plans to complete work to allow the *Employer* and Others to do their work and dates when the *Contractor* will need possession/access, acceptances and things provided by the *Employer*) become more important if a compensation event arises.

Because one of the important aspects of the programme is that it provides forensic evidence, the *Contractor* can use the programme to prove delays. He can only do this, however, if the dates when he required things were included in the programme.

> The programme can help both the *Employer* and the *Contractor* in assessing compensation events.
>
> The programme enables the *Employer* to carry out what-if scenarios.

Several compensation events relate back to the programme and the information that the *Contractor* has included in the programme. For example, the following are compensation events.

(1)   The *Employer* does not give possession/access by the later of a *possession/access date* and the date required by the Accepted Programme.[74] This relates back to the programme requirements for the *Contractor* to show:

- *possession/access dates* (the date when the *Employer* is allowing the *Contractor* access to and use of the Site) and
- the dates when he will need possession/access of a Site if it is later than the *possession/access date*.

Since the *possession/access dates* are included in the Contract Data, the non-inclusion of those dates in the programme will not necessarily result in the *Contractor* being denied this compensation event. Similarly, if possession is required later than the *possession/access date*, non-inclusion of this date will also not necessarily affect the *Contractor*'s right to the compensation event since the trigger could still be the *possession/access date* as stated in the Contract Data.

(2)   The *Employer* does not provide something, which he is to provide by the date for providing it required by the Accepted Programme.[75] This relates back to the programme requirements for the *Contractor* to show:

- the dates when the *Contractor* will need Plant and Materials and other things to be provided by the *Employer*.

The obvious question to ask is what happens if the *Employer* is to provide things to the *Contractor* but the *Contractor* has failed to include these elements in the programme? Clearly, if the *Employer* fulfils his obligations, there will be no repercussions. If, however, the *Employer* is late in his provision of an item, and the *Contractor* wishes to claim compensation, the event is not a compensation event described because the *Contractor* has not fulfilled his obligations in including the dates in his programme.

[74] Clause 60.1(2).
[75] Clause 60.1(3).

It appears that such neglect on the part of the *Contractor* is not envisaged by the contract, since the contract probably expects the *Contractor* to do whatever is within his power to complete the contract successfully. It may be interpreted that the *Contractor* has no recourse under the contract. However, it is likely that the *Employer*'s failure and the *Contractor*'s failure do not make it right to deprive the *Contractor*, when it would not be justified.

The *Contractor* may be able to produce a forensic analysis of his programme, working backwards in time to show when he would have expected the *Employer*'s item in order to carry out succeeding operations. However, it would be much simpler, less expensive, more effective and more cooperative just to include the information and dates on the programme.

(3)   The *Employer* or Others do not work within the times shown on the Accepted Programme.[76] This relates back to the programme requirements for the *Contractor* to show:

- The order and timing of the work of the *Employer* or Others either as stated in the Works Information or as later agreed with them by the *Contractor*.

The comments for the previous compensation event apply here as well.

All these elements of the programme emphasise the importance of the programme for the *Contractor* and for the mutual operations of the Parties. The proper and full drafting of the programme affects the smooth-running of the project and the *Contractor*'s ability to notify compensation events and maintain his profit.

> The programme is an essential part of assessing compensation events.

2.6.3.10 Method statements[77]   When the contract talks about the programme, it includes method statements; therefore each activity which may have several operations will require several method statements.

It should also be noted that (with Options A and C) the activities on the *activity schedule* should relate to the operations of each programme.

> Method statements (ECC3 'a statement of how the *Contractor* plans to do the work') are part of the programme.

When to call for method statements[78] and the like

Section 3 of the core clauses of the *conditions of contract* requires the *Contractor* to submit a great deal of information. It then becomes a management exercise to decide when to call for the information, which supports the Accepted Programme, taking into account the main Option applicable to the contract.

On large projects it may be impractical to call for all the method statements at the beginning of the project and a systematic approach to the submission and acceptance of method statements should be set up which allows adequate time for all the documentation to go through the *Employer*'s and *Contractor*'s quality systems.

On small projects it might be quite feasible to have all the information required submitted at the beginning of the project, perhaps prior to work commencing on Site.

---

[76] Clause 60.1(5).
[77] ECC3 – statement of how.
[78] ECC3 – statement of how.

2.6.3.11 Resource statement
The resource statement is something other than a method statement (which is also required with the programme) although it could be included in it. The resource statement is a description of the resources that the *Contractor* intends to use for each activity, such as labour and Equipment. The resource statement (and the method statement) is a part of the programme and is incorporated into the programme and is therefore required with every programme submitted, including those submitted as part of a quotation for a compensation event.

A resource statement could help the *Contractor* in his pricing, particularly if Option A (which uses an *activity schedule*) is the chosen pricing mechanism. Including the method and the resources required for each operation also allows the *Contractor* to plan his resources effectively and to be proactive in his procurement of those resources. Lastly, the effect of a compensation event on the programme is more easily assessed since the *Contractor* can immediately see the impact of any delays on his programme. If the *Project Manager* assesses the programme,[79] he will use those tools available to him, including the *Contractor*'s programme. If the programme is not up to date, the *Project Manager* may assess the compensation event based entirely on his own experience. The *Contractor* is thus incentivised to keep his programme up to date and accurate.

For operation 300(i), a resource statement might read as follows:

Activity No.	Description
300	Foundations up to DPC Level

Schedule of Equipment and other resources

- JCT 3XC driver
- 3T dumper driver
- Excavation gang
- 1 ganger
- 2 labourers

2.6.3.12 Float
It should be noted that the *Contractor* is required to show float and time risk allowances. Once these elements have been accepted by the *Project Manager* and incorporated into the Accepted Programme, they would tend to remain for the duration of the project.

There are a number of different types of float in programming. There is 'free' float between non-critical activities, 'total float' which is float on an activity and 'terminal float' which is any float that exists between planned Completion and the Completion Date.

- Free float — The amount of time a task/operation can be delayed before affecting any other task.
- Total float — The amount of time a task can be delayed without affecting planned Completion or reducing the terminal float.[80]
- Terminal float — Float attached to the whole programme and to sectional Completion (i.e. any float between planned Completion and the Completion Date).

The ECC clears up the old argument about 'entitlement' in that the terminal float belongs to the *Contractor*. Free float and total float are available to accommodate the time effects of compensation events in order to mitigate or avoid any delay to planned Completion and could therefore be said to belong to the *Employer*. **Time entitlement is also based upon entitlement and not need** as it is with traditional contracts, such as the ICE and JCT Contracts.

> The *Contractor* owns the terminal float.
>
> The assessment of time is based upon entitlement, not need.

[79]Clause 64.
[80]Where this has been examined by the *Project Manager* and accepted as realistic.

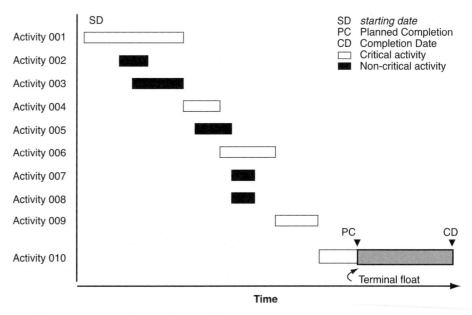

Fig. 2.3. Assessment of time effects of change

Therefore any compensation event which delays any activity on the critical path (i.e. an activity with zero float) must as a consequence delay planned Completion, giving rise to a commensurate delay to the Completion Date.

Let us run through an example. Figure 2.3 shows an outline programme for a project. The critical activities are shown by open boxes, the non-critical by solid boxes.

It can be seen from the diagram that the time between the planned Completion and the Completion Date is called the 'terminal float' and this float belongs to the *Contractor*.

2.6.3.13 Time risk allowances   The *Contractor* is required to show time risk allowances on his programme. Once they have been accepted by the *Project Manager* and incorporated into the Accepted Programme, they would tend to remain for the duration of the project. Time risk allowances are owned by the *Contractor* as part of his realistic planning[81] to cover his risk.

An example of this is the allowance he builds into his programme for down time occurring due to undertaking earthworks during the winter. The allowances should either be clearly identified in the programme as allowances, or be included in the time periods allocated to the specific activities.

2.6.3.14 Health and safety requirement   Some *Employers* might have additional health and safety requirements* that impact on the *Contractor*'s programme. These kinds of requirements should be clearly laid out in the Works Information and incorporated by the *Contractor* into his programme.

The *Contractor* should be able to demonstrate any allowances he has made in his programme for health and safety requirements. An example of this is time allowances for workmen to prepare themselves at the beginning of a shift and for washing themselves down and cleaning equipment when removing blue asbestos, which necessitates 'decontamination zones', 'sealed areas', etc.

With the advent of the CDM Regulations, this is a very important and serious consideration for the *Contractor*, who in the case of an accident might have to demonstrate to the Health and Safety Executive that he had allowed adequate time to undertake the work safely.

---

[81] Note the phraseology, which relates directly back to one of the *Project Manager*'s reasons for non-acceptance of the programme in clause 31.3.
*ECC2 clause 18.1; ECC3 clause 27.4.

The designers should stipulate in the document all the **job-specific** health and safety related works, for example confined space working and cleaning fuel tanks.

In some instances it may be difficult to show the allowances made in the programme. However, it is not impossible. Depending on the type of planning software used and the competence of the user of the planning tool, planners can produce programmes which can be very detailed, showing time/risk allowances, float, critical path, etc. These programmes can then be rolled up into a higher-level programme.

Where the use of the more sophisticated planning tools are felt not to be appropriate then in these instances the contract intends that these allowances should be shown in the method statements[82] for the work and which are part of the programme in the contract.

2.6.3.15 Procedures set out in the contract

This is a bit woolly for the ECC but is probably unavoidable in that the contract cannot cover all the possible alternatives. It is therefore very important for the drafters of the documents to ensure that they clearly and concisely spell out any other special requirements that they have in the Works Information.

It is also highlighting the need to recognise such things as the *period for reply* and is consequently warning the *Contractor* to allow for such things in the programme as the period the *Project Manager* has for accepting *Contractor* designs.

2.6.3.16 Other information

Other information is a very broad statement and it is important that the *Employer* states in the Works Information any information that he wants to see, bearing in mind that Works Information is not only the information included in the said document but could also be an instruction given to the *Contractor* in accordance with the contract.[83]

For example, this could cover:

- dates when, in order to Provide the Works, the *Contractor* is planning to obtain consents or
- dates when he is planning to submit any design to the *Project Manager* for acceptance.

This information can be useful to the *Project Manager* in planning when he needs particular resources available, for example designers.

It could also cover the dates when the *Contractor* needs completed Works Information. This will cover situations where the *Employer* has been unable to complete all the Works Information required by the *Contractor* in order for him to Provide the Works. This type of situation might occur where the *Employer* wishes to start a project quickly and has been unable to complete certain elements of the Works Information – items such as consents or approvals. In such instances the *Employer* should include assumptions within the Works Information.

Information release schedules should be included in the Works Information and the *Contractor* should be required to reflect in his first programme for acceptance the dates contained on the schedule. This way, as long as the *Employer* meets the dates, there will be no case for a delay to the Completion Date (unless of course the information is **different** from that assumed).

2.6.3.17 Actual progress

In addition to the items required by clause 31, clause 32 requires the *Contractor* to include other items into his revised programme, such as actual progress and its effect and the effect of implemented compensation events, and notified early warning matters, delays and notified Defects. In this way, the *Project Manager* is aware of whether the project is on time and, to a certain extent, on budget, and the *Contractor* is able to plan his programme and resources based

---

[82]ECC3 – statement of how the *Contractor* is to do the work.
[83]Clause 11.2(5) – ECC3 clause 11.2(19).

around contractual events. The Accepted Programme forms the as-built programme as time goes by, so that forensic analysis is facilitated through the resources and timing available on the programme.

2.7 Submission of the first programme

The ECC requires a first programme to be submitted by the *Contractor* to the *Project Manager* for acceptance either at tender stage[84] or shortly after the start of the contract.[85] Once the programme has been accepted, it becomes the Accepted Programme and both Parties work from this programme until the next programme is accepted and becomes the Accepted Programme.

> The first programme is submitted with the tender or within a stipulated time (e.g. four weeks) after contract award.

If the *Contractor* does not submit his first programme within the time required, the *Project Manager* retains 25%[86] of the Price for Work Done to Date to the *Contractor* until the first programme has been submitted.[87] This is clearly a powerful incentive to the *Contractor* and emphasises the importance attached to the programme by the ECC. Many employers prefer to see a programme with the tender submitted.

> The *Project Manager* may withhold a quarter of monies due (Price For Work Done To Date) if the *Contractor* does not submit a first programme.

A further note regarding the retention of the amount is that, as with other aspects of payment, such as delay damages, the retention of monies in this vein is not an option that may be exercised by the *Employer*. It is an obligation of the *Project Manager* to carry out his actions under the contract and to retain the 25%. He is not given the choice by the use of the word 'may'.

For some of the main Options in the ECC, the instruction to submit a programme with the tender may be pointless, particularly where information is minimal or where both Parties know that the information will be changing as drawings are amended or as other information is revealed. Some companies still refer to a programme submitted with the tender as the 'tender programme' and a revised programme submitted after the contract as the 'contract programme'. There is no place in the ECC for this kind of terminology. In addition, it tends to imply that the tender programme is of less importance than the contract programme, whereas in the ECC, once accepted, each programme is important.

2.8 How often is the programme revised?

The *Contractor* submits a revised programme to the *Project Manager* for acceptance on four different occasions:

(1) **Every regular period during the contract**[88]
Contract Data part one dictates the period within which the programme has to be revised. This could be four weeks or eight weeks, or even three months depending on the complexity and length of the project. Some employers even ask for two different types of programme, one of which is more complex and is to be submitted less frequently. Some employers also require a summary programme at weekly progress

[84] Where the optional statement is included in Contract Data part two.
[85] Where the optional statement is included in Contract Data part one.
[86] Clause 50.3.
[87] Note that it is the submission of the programme that is important, not the acceptance of the programme by the *Project Manager*.
[88] Clause 32.2.

meetings. Although this will clearly be based on the Accepted Programme, its requirement is more likely to be described in the Works Information.

> The *Contractor* updates the programme regularly.

It is recommended that employers take heed of the complexity and duration of the project in deciding how frequently they wish to see the programme. Although the programme is a pivotal project management tool, it can also be an onerous and possibly expensive document for the *Contractor* to produce, and employers should try to avoid making the task more difficult for contractors in requesting very frequent revisions unnecessarily.

(2) **If he chooses to do so**[89]

The *Contractor* may choose to submit a programme for acceptance to the *Project Manager* outwith the obligatory regular period. An example is where he has amended sequencing or he has changed the method or resourcing of an activity. The *Project Manager* is obliged to reply to the *Contractor* with his acceptance or otherwise of the programme within two weeks of the *Contractor* having submitted it.

(3) **When he has been instructed to do so by the *Project Manager***[90]

If the *Project Manager* instructs the *Contractor* to submit a revised programme for acceptance, the *Contractor* is obliged to submit the programme within the *period for reply*.

(4) **If a compensation event has affected the programme**[91]

A quotation for a compensation event comprises changes to the Prices and a delay to the Completion Date. Where the compensation event has had (or will have) the effect of changing the programme, then the *Contractor* has to submit a revised programme with his quotation showing the effect of the event (or, in ECC3, the alterations to the Accepted Programme). If the programme is affected simply because a method[92] or resourcing changes, but the Completion Date remains the same, a revised programme still has to be submitted since both the resources and the method are a part of the programme.

The time periods for the submission are as described in the compensation event procedure, but in general, the *Contractor* has three weeks to submit a quotation from being instructed to do so. Note that if the *Contractor* does not submit a revised programme with a compensation event quotation as required, the *Project Manager* may make his own assessment of the compensation event.

2.9 Acceptance of the programme

The *Project Manager* may accept or not accept the first programme and each subsequent programme submitted for acceptance by the *Contractor* based on various criteria stated in the contract. It is only when a programme is accepted that it becomes the Accepted Programme.

> If the *Project Manager* does not accept the programme for any reason other than the four stated in clause 31.3, it is a compensation event.

Note that if there is no Accepted Programme – that is, if the first programme submitted by the *Contractor* has not been accepted – the *Project Manager* may

---

[89] Clause 32.2.
[90] Clause 32.2.
[91] Clause 62.2.
[92] ECC3 clause 31.2 – how the *Contractor* plans to do the work.

use his own assessment of the programme for work that is affected by a compensation event.[93] The *Project Manager* may do the same if the *Contractor* has not submitted a revised programme (or, in ECC3, alterations to a programme) for acceptance as required by the contract.

2.9.1 Reasons for non-acceptance

There are only four reasons for the *Project Manager* to refuse acceptance of the programme. They are as follows:[94]

(1)   the *Contractor*'s plans which it shows are not practicable,
(2)   it does not show the information which this contract requires,
(3)   it does not represent the *Contractor*'s plans realistically or
(4)   it does not comply with the Works Information.

Although in general the clauses of the contract are very clear, it would be fair to say that one or two of the above reasons for refusing acceptance of a programme could be interpreted in various ways. In particular, realistic or practicable planning tends to be subjective and therefore any non-acceptances based on these reasons should be carefully considered. The following subsections discuss the reasons in more detail.

2.9.1.1 The Contractor's *plans, which it shows, are not practicable*

The first reason refers to the *Contractor*'s plans only. An example is where the *Contractor*'s programme shows his planned progress on a tunnel to be 90 m/week. His original accepted programme shows him achieving 10 m/week and his actual rate of progress is 50 m/week. You are also aware that the best output ever achieved by a *Contractor* on a tunnel of this type under similar conditions is 75 m/week. His plans are therefore not realistic.

When assessing tender submissions, data relating to output rates/production rates may be very useful to benchmark the respective tenders received.

2.9.1.2 It does not show the information, which this contract requires

The second reason refers to the contract and it should be noted that references to the contract include not only the ECC *conditions of contract* but also the Works Information and whatever other information and documentation has been incorporated into the contract. An example of this is where the programme does not show the *possession/access date* or sectional Completion Dates.

Can the *Project Manager* refuse to accept a programme because the Completion Date shown on the submitted programme is later than the Completion Date shown on the Accepted Programme, but there is no compensation event or other reason for the date to have been delayed? One could argue that the reason for non-acceptance 'does not show information that this contract requires' could apply since the contract requires the Completion Date to be shown, whereas the date shown as the Completion Date on the programme is not the Completion Date. The same reason could be argued where the actual progress has not been measured accurately against planned progress in the programme.

2.9.1.3 It does not represent the Contractor's *plans realistically*

An example of the third reason is where a *Contractor*'s programme assumes use of driven piles and the *Project Manager* is aware that the site team is using bored piles.

Another example is where a *Contractor*'s programme shows commissioning/testing a bank of six lifts at the same time, this requiring **all** the UK suppliers' commissioning engineers to be used at one time.

2.9.1.4 It does not comply with the Works Information

An example of the fourth reason is that the programme has not taken into account any design constraints stated in the Works Information. An illustration of this could be where the *Contractor* cannot undertake a certain item of the *works* until certain other things have been completed.

---

[93] Clause 64.2.
[94] Clause 31.3.

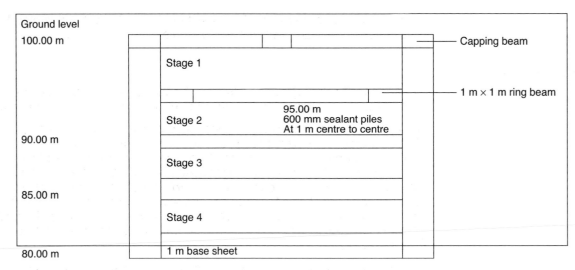

Fig. 2.4. Secant piled lift shaft box

Let us say that a new secant piled lift shaft box is to be constructed top down (see Fig. 2.4). The Works Information might stipulate that the lift shaft box is to be constructed top down in a series of stages as follows:

Stage 1: level 100.00 to 95.00 m
Stage 2: level 95.00 to 90.00 m
Stage 3: level 90.00 to 85.00 m
Stage 4: level 85.00 to 80.00 m

No excavation work can commence until the previous stage is complete and the 1×1 m concrete ring beam at each level has been left to cure for a minimum of 14 days after the last concrete pour.

If the *Contractor* shows an activity on his programme allowing only seven days for curing before excavating the next level, then the work is not in accordance with the Works Information.

2.9.2 Resubmission of an unaccepted programme

If the *Project Manager* does not accept a programme, the *Contractor* is obliged to resubmit the programme within the *period for reply*.[95] In other words, the *Project Manager* does not have to instruct the *Contractor* to submit a revised programme because the *Contractor* is obliged to do so anyway under the *conditions of contract*. Once the *Contractor* receives notification of non-acceptance, he should automatically set about drafting a revised programme for submission to the *Project Manager*.

2.9.3 Timing of the acceptance or non-acceptance

The *Project Manager* has two weeks to reply to the *Contractor* after the *Contractor* submitted his programme for acceptance. As with so many other clauses in the contract, the ECC does not necessarily envisage the *Project Manager* not carrying out his actions under the contract and there is no default position stated in the ECC where the *Project Manager* does not reply within his allotted time period.

It is not clear therefore whether a non-reply is deemed to be non-acceptance or whether it is deemed to be acceptance. In practice, it is possible that it may be the latter in order not to prejudice the *Contractor*'s position.

In addition, it is a compensation event[96] if the *Project Manager* does not reply to a communication within the time period stated in the contract. The *Contractor* therefore becomes entitled to an assessment of time and money for the

[95]Clause 13.4.
[96]Clause 60.1(6).

Project Manager's tardiness. This sanction for a late reply reinforces the view that a non-reply is deemed to be acceptance.

2.10 Completion

The *Contractor* is required to Provide the Works so that Completion is on or before the Completion Date.[97] Time is therefore of the essence in ECC contracts.

Completion takes place when the *Project Manager* decides that Completion has been achieved.[98] This could be on, before or after the Completion Date. The *Contractor* is not required to notify the *Project Manager* that he considers the *works* to be complete. The *Project Manager* need not take heed of any such notice that the *Contractor* chooses to submit. The *Project Manager* would clearly take into account the details in the Works Information about when Completion takes place. In addition, Completion cannot take place if there are still uncorrected Defects that would prevent the *Employer* from using the *works* and Others from doing their work.[99]

The *Project Manager* is required to certify Completion within one week after he decides that Completion has taken place.

2.10.1 Take over and Completion Take over and Completion are linked by time and circumstance. Take over generally follows Completion; however, sometimes take over could occur before Completion. The ECC details the procedures to be followed in all cases.

2.10.1.1 Take over before Completion If the *Employer* starts to use any part of the *works* before Completion has been certified, then he is deemed to have taken over the *works*. The exception to this is where the take over has occurred because of a reason stated in the Works Information or to suit the *Contractor*'s method of working.[100]

2.10.1.2 Take over after Completion The *Employer* takes over the *works* not more than two weeks after Completion.[101]

Note that the *Employer* need not take over the *works* before the Completion Date if he has stated in the Contract Data that he is unwilling to do so. This is different from taking over before Completion. The former case refers to the situation where the *Contractor* has completed the *works* before the Completion Date stated in the contract, but the *Employer* is not willing to take over the *works* earlier than he would have if the *Contractor* had completed on time.

2.11 Take over by the *Employer*

2.11.1 How it happens Taking over is the term used by the ECC to signify that the *Employer* has taken over the *works* or a part thereof.

The *Project Manager* certifies take over of any part of the *works* within one week of the date of take over.[102]

2.11.2 What it means Take over means that possession/access of the site or part thereof returns to the *Employer*. The *Contractor* no longer has a right to access and the *Employer* has to make arrangement for access for the *Contractor* in order for him to correct Defects which occur.

It also signifies a risk transfer in terms of liability for loss/damage to the *works* from the *Contractor* to the *Employer*.

---

[97] Clause 30.1.
[98] Clause 30.2.
[99] ECC3 clause 11.2(2); ECC2 clause 11.2(13).
[100] ECC2 clause 35.3 – ECC3 clause 35.2.
[101] ECC2 clause 35.2 – ECC3 clause 35.1.
[102] ECC2 clause 35.4 – ECC3 clause 35.3.

2.12 Acceleration

Acceleration under the ECC deserves a special mention because it is treated differently from traditional contracts.

Acceleration is described in the contract as being an instruction from the *Project Manager* to achieve Completion before the Completion Date. This description of acceleration is not requesting the *Contractor* to speed up his progress in order to achieve Completion on the Completion Date, which is something else entirely.

Acceleration is a useful provision for those clients for whom time is critical. For example, retail companies have critical times such as Easter and Christmas, on which they might wish or need to be open.

The price to the *Employer* for achieving acceleration is whatever the *Contractor* chooses to quote to the *Employer*.[103] Acceleration is not a compensation event and therefore the basis of the quotation does not have to be the Actual/Defined Cost. The *Contractor* does not even have to submit a quotation as long as he gives his reasons for not doing so. Acceleration is therefore purely in the *Contractor*'s control. Of course, most reasonable contractors are likely to give a fair and reasonable quotation assuming that mutual trust and cooperation is working in the contract.

> The *Contractor* is not obliged to accelerate.

There is an alternative way of maintaining the Completion Date when changes occur to affect the programme. In the compensation event procedure, the *Project Manager* may instruct the *Contractor* to submit alternative quotations based on different ways of dealing with the compensation event.[104] If the *Project Manager* wishes to maintain the Completion Date despite a compensation event that could have delayed the Completion Date, he could instruct the *Contractor* to submit an alternative quotation maintaining the Completion Date. This could involve overtime or a change to the sequencing of the programme or both and is likely to cost more than a simple change to the Prices combined with a delay to the Completion Date. In any case, the *Project Manager* then has the choice of quotations.

2.13 Other aspects of programming in the ECC

2.13.1 *Contractor*'s programme

It is important to understand that the programme is the *Contractor*'s programme. The *Project Manager* does not have the same powers to amend the programme as in traditional contracts. There is no statement in the ECC that the *Contractor* shall use best endeavours to progress the *works*, nor does the *Project Manager* have the ability to instruct the *Contractor* to speed up his progress to ensure that Completion is on or before the Completion Date.

If the *Contractor*'s actual progress is lagging behind his planned progress, there is not much that the *Project Manager* can do to encourage the *Contractor* to complete on time other than instruct the submission of another programme. This probably works on the assumption of mutual trust and cooperation and also that it is not really in the *Contractor*'s interests to work slowly since (depending on the payment option) this could cost the *Contractor* money.

The *Project Manager* has the choice of not accepting a subsequent programme if it does not show the *Contractor*'s plans realistically (where the programme still shows progress on schedule), but this does not necessarily achieve what the *Project Manager* really wants, which is for the *Contractor* to speed up.[105]

---

[103] Clause 36.1.
[104] Clause 62.1 – ECC3 makes it a requirement of this clause that, 'After discussing with the *Contractor* different ways of dealing with the compensation event which are practicable.'
[105] As opposed to accelerate.

Some employers therefore include in the contract that the *Project Manager* may instruct the *Contractor* to speed up his progress (and through clause 29.1, the *Contractor* has to obey). If the payment Option is C, D or E, the *Employer* might even include the cost of this speeding up as a Disallowed Cost.[106]

2.13.2 *Employer's* programme

Some employers call the master plan or schedule that they have drafted the 'programme', and the contractor is expected to conform to the employer's programme.

This is all very well, but it is unlikely that the *Employer's* programme includes all the aspects of the programme as described above and which the *Contractor* is obliged to include in his programme. The *Contractor* might buy into the *Employer's* programme as the master programme for the overall project, but would be obliged under the contract to produce his own programme that would provide for his notifying compensation events and would provide the information that the *Project Manager* needs in order to manage the project effectively.

The above also applies to Subcontractors when working for *Contractors*.

2.13.3 Affecting cash flow

The programme may affect cash flow, particularly for Option A contracts, where the *Contractor* is not paid for an activity until he has completed that activity. The *Contractor* therefore needs to ensure that the activities described in the programme can be completed before an assessment date,[107] so that he (the *Contractor*) is always assured of completed activities and therefore payments.

The problem is that the shorter the activities are, the more opportunity there is for Completion of those activities and therefore payment, but the longer it takes to produce the programme. The *Contractor* should therefore try to reach a balance between the length of the activities and need for cash flow.

2.13.4 Moving target

The Accepted Programme as a management tool under the ECC is a moving target and as such needs to be regularly updated and reviewed. This review period will depend upon the size/scope of the project; on major projects it could be done weekly whereas on short-term fast-track projects, for example shop fit-outs, it may need to be daily.

There are tight time-scales to be met on its submission and acceptance and sending programme information backwards and forwards between the *Contractor's* and *Employer's* planners wastes valuable time.

To make it work to its full potential, an integrated programme is required and a planning team which monitors, reviews and agrees changes to the Accepted Programme on a continuous basis so that the submission of the programme as stated in the Contract Data becomes a mere formality.

The *Employer* could also ensure that the Works Information details the programme IT software requirements so that both sides are using the same software, facilitating quick and easy information transfer between them.

Because the programme is a moving target, the *Employer's* planner needs to constantly check the dates by which, in order for the *Contractor* to Provide the Works, the *Employer* is required to provide facilities, possession/access, Works Information, etc. It should be noted that the *Employer* has no obligation to provide 'things' earlier than dates stated in the Works Information.

If the *Contractor* should propose earlier dates than those shown for the *Employer* to provide information or possession etc. then the *Project Manager* should accept these earlier dates **if it is acceptable to the *Employer***. If it is not, and the *Project Manager* accepts a programme showing earlier dates, then failure to meet these earlier dates will give rise to a compensation event.

---

[106] By amending the Disallowed Cost definition in clause 11.2.
[107] See Chapter 2 section 2.4.1.3 in Book 2. Footnote: ECC3 clause 27.3.

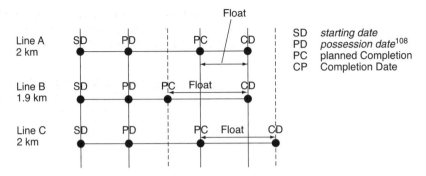

Fig. 2.5. Reducing and increasing the scope of the *works*

2.13.5 What happens if the scope of the *works* is reduced or increased?

If you had a contract to build 2 km of new road and it was decided to reduce this to 1.9 km then, all things being equal, planned Completion on the Accepted Programme would be earlier.

In Fig. 2.5, line A shows the original *starting date, possession/access date,* planned Completion and the Completion Date. In line B with the reduced scope it can be seen that planned Completion moves forward, increasing the terminal float which is owned by the *Contractor*. The Completion Date does not move.

If later it is decided to increase the scope of the *works* back to the original scope then it can be seen in line C that planned Completion reverts to the date shown on the original programme, but because the *Contractor* owns the increased float, the Completion Date moves by the same amount as the planned Completion Date.

This is another subtle message to the *Employer* to ensure that he knows exactly what it is that he intends to build.

2.13.6 What happens if the activities on the revised programme do not match the *activity schedule*?

The contract requires that the activities on the *activity schedule* **must relate to the programme** (Options A and C only). If they are not then there exists the potential problem whereby delays could occur to an activity that is listed on the *activity schedule* but that does not appear on the programme. At this point the programme fails to be a useful management tool.

If this occurs the *Project Manager* should not accept the revised programme assuming that the start and finish date of each activity are shown. He should ask the *Contractor* under clauses 32.2 and 54.2 to submit a revision of the *activity schedule* to the *Project Manager* for acceptance.

The *Contractor* is at a disadvantage at this stage because if the *Project Manager* has not accepted a revised programme, he can make his own assessment of a compensation event.

2.13.7 What happens if the *Contractor* fails to maintain an Accepted Programme?

In this instance the onus to maintain an up-to-date programme would seem to fall upon the *Project Manager*. It is not a strict contract requirement, but programme maintenance is necessary for the *Employer*'s right to levy delay damages otherwise it will be frustrated. The *Project Manager* will need to have the capability to manage the programme so that he can make *Project Manager*'s assessments for compensation events. In this instance the *Project Manager* can reject all quotations submitted by a *Contractor* because he has not maintained an Accepted Programme.

2.13.8 The importance of good site records

Good site records[109] are an essential tool in the management of time and it is essential that the *Supervisor* realises the importance of keeping good site diaries and how this information assists in the overall management of the contract.

---

[108] ECC3 – *access date.*
[109] See Chapter 1 of Book 5 for more information on site records.

For example, the site diary can be invaluable in establishing the events that took place on any particular day in the case of compensation events over causes of and responsibility for delay and/or disruption.

> Good site records can prove invaluable in establishing the facts and events when assessing compensation events.

3 Control of quality

Synopsis

This chapter discusses:

> - The quality framework embedded within the ECC
> - The *Contractor*'s obligations
> - Role of the *Employer*'s representatives
> - Subcontracting
> - Quality control

                                                          www.neccontract.com

3.1 Introduction

Construction contracts have traditionally had embedded within them a 'quality framework' providing the following features:

- rights of access for the employer or his representative,
- powers and duties given to the employer or his representative relating to quality control,
- powers for the employer or his representative to deal with defective work,
- powers of enforcement in case the *Contractor* does not respond.

This quality framework is the contractual approach to assuring quality. It is intended to establish quality requirements, ensure compliance and avoid defects as far as possible and to deal with defects without recourse to legal proceedings.

The latest extension of quality management principles in construction is a movement towards increasing reliance on inspection and testing by contractors, whether of their own work or of work produced by their subcontractors.

The rest of this chapter sets out to analyse how quality standards are set down in the ECC and to consider how comfortably this sits with the quality management principles and practices reflected in the current Quality Standards.[110]

3.2 The ECC and quality

It is intended to analyse the particular clauses constituting the quality framework contained within the ECC under the following main headings:

- *Contractor*'s obligations,
- role of the *Employer*'s representatives,
- *Employer*'s supply of Plant and Materials and Equipment,
- subcontracting,
- quality control,
- defective work,
- certification,
- enforcement.

Figure 3.1 shows these elements of control of quality on an ECC contract in diagrammatic form.

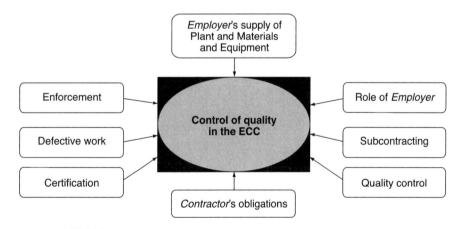

Fig. 3.1. Control of quality in the ECC

3.3 The *Contractor*'s obligations

3.3.1 General obligations   The *Contractor*'s general responsibility for quality is part of his wider duty to Provide the Works in accordance with the Works Information.[111] Expanding this

---

[110]BS 5750: 1979 Parts 1, 2 and 3 as International Standards ISO 9001, 9002 and 9003 and as European Standards EN 29001, 29002 and 29003.
[111]Clause 20.1.

---

duty in line with the defined terms, the *Contractor* is responsible for doing 'the work necessary to complete the *works* in accordance with this contract and all incidental work, services and actions which the contract requires'.[112]

The Works Information is part of the contract, its purpose being to specify and describe the *works* and to describe any constraints on how the *Contractor* Provides the Works. It follows therefore that the quality standards to be achieved by the *Contractor* should be specified in the Works Information prepared by or on behalf of the *Employer*. These standards are vitally important to quality control since they provide the basis upon which the existence of a Defect is judged.[113]

> The *Contractor*'s general responsibility is to Provide the Works in accordance with the Works Information.
>
> Quality standards to be achieved by the *Contractor* should be specified in the Works Information.

One of the objectives set by the NEC's drafting panel was that the contract should contribute to a reduction in the incidence of disputes. One of the improvements it sought to introduce in pursuit of this objective was clarity. It therefore follows, if the objectives of the contract are not to be undermined, that the quality standards described in the Works Information should be expressed so that compliance is capable of objective assessment. This approach is entirely compatible with the principles of quality management, which demand that the supplier (*Contractor* or Subcontractor) should be able to verify compliance directly without reference to the purchaser (*Employer* or his representative). Consequently as far as it is practicable, the traditional specifying of requirements by reference to the opinion of the *Employer* (or his Architect/Engineer/ *Supervisor*) should be eliminated from the Works Information.

> The quality standards described in the Works Information should be expressed so that compliance is capable of objective assessment.
>
> Specifying requirements by reference to the opinion of the *Project Manager* should be eliminated from the Works Information.

3.3.2 Ambiguities and inconsistencies (discrepancies) in or between the contract documents

A difficult area concerns identification by the *Contractor* of discrepancies in the contract documentation. As a general principle the economic achievement of quality is assisted by the identification of such shortcomings at the earliest possible stage of the construction process. The ECC is entirely consistent with this principle, placing upon both the *Contractor* and the *Project Manager* in clause 17.1 the obligation to give a notification as soon as one or other of them becomes aware of an ambiguity or inconsistency in or between the documents comprising the contract. The *Project Manager* then has the responsibility for instructing the resolution of the ambiguity or inconsistency.

> There is an obligation on the *Contractor* and *Project Manager* to give a notification as soon as they become aware of an ambiguity or inconsistency.
>
> The *Project Manager* resolves any ambiguities and inconsistencies between contractual documents by giving an instruction.

Provisions in some standard forms of contract for reimbursement to the *Contractor* because of such discrepancies in contract documentation act as a disincentive to the *Contractor* to comment before the contract is let, and a

---

[112]ECC2 clause 11.2(4); ECC3 clause 11.2(13).
[113]See section 3.7.3 below on tests and inspections.

possible disincentive to early reporting after the contract is let. Likewise, when using some standard forms of contract, which envisage the same organisation preparing the contact documentation and administering it, there may be reluctance on the part of individuals to draw attention to discrepancies in documentation, which their company prepared.

The ECC encourages both parties to notify as soon as they become aware of anything, which may affect cost, time or quality of the *works*. The *Contractor* is incentivised to notify as soon as he becomes aware of these matters. The sanction for failure to do so is that he runs the risk of losing his full entitlement to the time and cost consequences of the event.

The *Employer*'s sanction is that he could pay more time, more money and not get the quality that he was hoping for. For the *Project Manager* and *Supervisor* there is no hiding place and their actions/inactions will be highly visible.

A concept **not** found in the ECC is a hierarchy of the documents forming the contract. Such a hierarchy is used in some standard forms as a means of resolving ambiguities and discrepancies by giving precedence to documents higher up the contractual hierarchy. With one exception to this general rule (see below) the ECC contains no such hierarchy. Instead, very simply, having been notified of an ambiguity or inconsistency, the *Project Manager* resolves the problem by instructing a change to the Works Information. This of course is a compensation event and by the operation of clause 63.7[114] any assessment is based on the interpretation most favourable to the Party **not** responsible for the ambiguity/inconsistency; that is, the contra proferentem rule operates, which interprets a clause containing an ambiguity or inconsistency against the Party responsible for drafting the document in which it occurs.

> In ECC the concept of 'Key Dates' is added to 'the Prices' and 'the Completion Date' as a matter for early warning.

The one exception to the 'no-hierarchy' concept is where an inconsistency becomes apparent between the Works Information provided by the *Employer* and a design which is part of the Works Information provided by the *Contractor* and referenced in part two of the Contract Data. On the assumption, in this instance, that having been notified of the ambiguity/inconsistency, the *Project Manager*'s instruction required the design to comply with the *Employer*'s Works Information, such a change would **not** be a compensation event.[115] This effectively gives precedence to the Works Information provided by the *Employer* over the Works Information provided by the *Contractor*.

> There is no hierarchy of documents in the ECC.
>
> The contra proferentem rule applies to the Works Information.

3.3.3 *Contractor*'s design

There is no '*Contractor*-design' version of the ECC. That is not to say the contract does not provide for all or part of the *works* to be designed by the *Contractor*. On the contrary, it offers the flexibility of being suitable for both 'traditional' arrangements where the permanent *works* are designed by or on behalf of the *Employer* and for 'design-and-build' arrangements where the *Contractor* designs and builds the whole of the *works* in accordance with criteria set down by the *Employer*. Between these two extremes it is suitable for the many instances where the *Contractor* does some design and some is done by, or on behalf of, the *Employer*.

But how is this flexibility achieved? Very simply, the ECC requires the Works Information to state the parts of the *works* which the *Contractor* is to design and the criteria to which such designs are required to conform. Such criteria

---

[114]ECC3 clause 63.8.
[115]See second bullet point of clause 60.1(1).

could include details of the form, geometry, dimensions, specifications, codes of practice, standards and environmental criteria, or alternatively could comprise a performance specification where the *Contractor* is responsible for the majority of the design. If, after having let the contract, it is decided to increase or decrease the areas of the *works* that the *Contractor* is to design, or to change the criteria to which he is already designing, such changes would be in the form of instructions changing the Works Information and would consequently be a compensation event.

> The Works Information should state the parts of the *works* which the *Contractor* is to design.

3.3.4 Supervision/employees

Contrary to many quality assurance practitioners' belief, quality is not achieved by procedures, it is achieved by people. Procedures do not take decisions, people do.

You could have the best procedures in the world and still not achieve quality – there has to be a combination of both people and procedures to achieve the desired quality.

Neither the traditional standard forms of contract nor the published quality standards have placed enough emphasis on the competence (quality) of people. Competency is a function of knowledge and skills gained from education, training and/or experience.

The ECC[116] places an obligation on the *Contractor* to employ the key persons named by the *Contractor* in part two of the Contract Data to do the jobs designated to them therein. If maximum value is to be gained from this feature of the ECC the *Employer* should indicate in the instructions to tenderers the key persons for whom details (including job description, responsibilities, qualifications and experience) are required. Such information when received with tenders should then be assessed as part of the wider tender evaluation.[117]

Clause 24.1 also requires that if the *Contractor* wishes to provide replacement people, he must provide people of equal quality and ability. This feature has been introduced so that the *Contractor* provides the required quality of people, for example the 'A' team as outlined in their tender submission. Very often in the past, contracts were awarded on the basis of the CVs put forward, only for the *Employer* to find that these people did not put in an appearance on the contract.

> People achieve quality.
>
> The *Employer* should indicate in the instructions to tenderers the key persons for whom details are required.

The ECC in clause 24.2 also recognises that the *Project Manager* should be able to have an employee removed from the project.

3.3.5 Mode and method of construction

It is often overlooked that any reference in the ECC to the programme includes method statements. Clause 31.2 requires the *Contractor* to include with each programme which he submits 'a method statement which identifies the Equipment and other resources which the *Contractor* plans to use' for each operation.[118]

---

[116] Clause 24.1.

[117] See Appendix 1 to Chapter 1 of Book 2.

[118] In ECC3 this requirement has been deleted and instead requires 'for each operation a statement of how the *Contractor* plans to do the work identifying the principal Equipment and other resources which he plans to use'.

> Any reference in ECC2 to the programme includes method statements. In ECC3 clause 31.2 requires the *Contractor* to provide a statement of how he plans to do the work.

The traditional reason for asking for method statements, particularly in civil engineering contracts, is to allow the designer of the permanent works (traditionally the Engineer) to check that the proposed methods of construction will not have a detrimental effect on any partly completed permanent *works*. It is often overlooked by the drafters of ECC-related contract documentation that, unless the Works Information specifically precludes particular methods of construction, any constraints subsequently introduced would under the ECC constitute a change to the Works Information and would consequently be a compensation event. This is particularly relevant on *works* involving heavy foundation engineering where the trend is to reduce the need for (and accordingly the cost of) temporary earthwork support systems by exploiting the ability of the permanent structure to provide such support during construction. This applies notably to the technique of 'top-down' construction and the use of embedded peripheral walls (e.g. secant/contiguous piled walls and diaphragm walls).

A more recent development that has led to a clamour by employers for method statements is the publication in 1992 of the Management of Health and Safety at Work Regulations. Under these regulations if a *Contractor* is to carry out work which has risks to health and safety (and most construction activities do), he has a legal duty to carry out a risk assessment. Although not required by law, preparing a written method statement after carrying out a risk assessment has proved to be an effective way of producing an action plan identifying the necessary health and safety measures to be employed to control the risks identified. If it is likely that the *Project Manager* will want to see safety method statements (as evidence that the required risk assessments have been undertaken) then it is probably wise to include a statement to this effect in the Works Information describing the health and safety requirements.[119]

It is worth re-emphasising that the ECC treats method statements submitted by the *Contractor* as part of the programme and consequently the reasons for not accepting the programme apply equally to these method statements, the reasons being that the method statements:

- are not practicable,
- do not show the information which the contract requires (i.e. they do not identify the Equipment and other resources which the *Contractor* plans to use),
- do not represent the *Contractor*'s plans realistically,
- do not comply with the Works Information, for example they may show that the *Contractor* plans to employ a method of construction that is expressly precluded by the Works Information.

Acceptance of a programme (and therefore method statements), unlike acceptance of the *Contractor*'s design, is not a condition precedent to the *Contractor* proceeding with the work. Failure by the *Contractor* to submit at the required times, or a decision by the *Project Manager* not to accept a revised programme (and therefore revised method statements), does not require the *Contractor* to stop work. A lack of attention however by the *Contractor* to his obligations regarding programmes and method statements does under the ECC expose the *Contractor* to the risk of the *Project Manager* carrying out his own assessments of any compensation events that arise.[120]

[119] Refer to clause 18.1. In ECC3 clause 18.1 is deleted and is replaced as clause 27.4.
[120] Refer to clause 64.1 third and fourth bullet points.

3.3.6 Setting out Unlike most other standard forms of contract, the ECC does not contain an express condition making the *Contractor* responsible for the setting out of the *works*. It is arguable that the definition of 'to Provide the Works'[121] is sufficiently wide to include setting out since it includes for all incidental works and services. Since the *Contractor* is required to Provide the Works 'in accordance with the Works Information' (in clause 20.1 describing the *Contractor*'s primary obligation), it is recommended that the Works Information makes it clear that the setting out of the *works* is the *Contractor*'s responsibility.

> The Works Information needs to include a statement with regard to the *Contractor*'s responsibility for setting out the *works*.

3.3.7 Quality management systems The ECC does not include a requirement for the *Contractor* to establish and maintain a documented quality management system. The ECC2 Guidance Notes state:

> 'Requirements in the ECC for quality systems can be accommodated in two ways as follows:
>
> (a) The *Employer* specifies requirements for quality management procedures in the Works Information;
> (b) The *Employer* requires the *Contractor* to provide details of his quality plan in the Works Information.'[122]

It is contended that both of these methods on their own, without any reference to the status of either the quality management procedures or the quality plan in the conditions, are unsatisfactory and that it is both sensible and necessary to introduce changes to the ECC, such as to add obligations in section 4 of the core clauses for the *Contractor*:

- to operate a quality management system complying with the requirements stated in the Works Information,
- to submit to the *Project Manager* for acceptance (within a stated period) a quality policy statement and a quality plan which include information required by the Works Information,
- to comply immediately with any instruction from the *Project Manager* to correct a non-compliance with any quality management system by the *Contractor*.

With regard to these potential additions to the ECC the following observations are made:

(1) Unless the Works Information describes both the requirements for the quality management system and states the information which the quality policy statement and quality plans are to include, the additional clauses will be of limited effect.

(2) Three reasons could be included as to why the *Project Manager* may not accept the *Contractor*'s quality policy statement and his quality plan as follows:

- 'they are not practicable' (meaning that the *Project Manager* does not consider that if implemented they will assist in achieving the desired standards set out in the Works Information),
- 'they do not show the information which this contract requires',
- 'they do not comply with the Works Information' (most likely reason provided the Works Information has been prepared thoroughly).

(3) As with all other instances where the *Project Manager*'s acceptance is required, if the latter withholds acceptance for a reason other than one stated in the contract, a compensation event arises.[123]

---

[121]ECC2 clause 11.2(4); ECC3 clause 11.2(13).
[122]Page 43, Quality Systems paragraphs 1 and 2.
[123]Clause 60.1(9).

(4) Acceptance by the *Project Manager* of the *Contractor*'s quality policy statement and quality plan is not intended to change the *Contractor*'s liabilities or his responsibilities to Provide the Works.[124] This is very important with respect to quality plans: there is no guarantee that the end-product requirements for the *works* (as described in the Works Information) would be met just because the *Contractor* faithfully adheres to the documented procedures and instructions.

(5) The period stated (calculated from the *starting date*) for submission by the *Contractor* of the quality policy statement and quality plan should be mindful of the need to have these documents in place before any serious work is commenced. To speed up the process it may be worth considering asking *Contractor*s to include with their tender submissions an overall outline quality plan and a detailed quality plan covering the first six weeks' activities.[125]

(6) The Works Information should set out the degree of involvement of the *Project Manager* in terms of the latter's 'external supervision' of the *Contractor*'s quality management system (e.g. external audits, hold and witness points expressly required).

(7) Non-compliance with the quality system, such as a failure to keep records, does not necessarily cause any material loss or damage or give rise to a Defect, but confidence in the system demands full compliance. Consequently, the wording of any 'enforcement' clause has to be carefully thought through and drafted. Such a clause could require the *Contractor* to comply with an instruction from the *Project Manager* 'to correct a failure to comply with the quality plan'. This is sensible since at the time that a non-compliance with the quality plan is identified there may be no evidence of a Defect as defined by the contract (clause 11.2(15)).[126] Of course if the *Project Manager* considers that the nature of the non-compliance is highly likely to have given rise to a Defect, he has the power to instruct a search.[127]

3.4 Role of the *Employer*'s representatives with respect to quality

The ECC envisages two '*Employer*'s representatives', the *Project Manager* and the *Supervisor*.

The *Project Manager*'s role is to manage the contract on behalf of the *Employer* with the intention of achieving the *Employer*'s objectives (usually expressed in terms of a budget, a programme and a brief setting out the requirements for the end-product). His authority under the contract is expressed in terms of the actions that the contract prescribes to him and includes authority to change the Works Information, to instruct the *Contractor* to do various things and to generally exercise his managerial and engineering judgement.

The *Supervisor* has a much more restricted role, effectively limited to ensuring the *works* are constructed in accordance with the Works Information. The importance of the role should not be underestimated given its obvious connection with one of the *Employer*'s objectives, namely to ensure that the end-product meets the requirements set out in the client's brief. The ECC Guidance Notes do not explain why the drafters of the contract considered it necessary to split duties between the *Project Manager* and the *Supervisor* although it is likely that they considered there to be a potential conflict between getting the *works* constructed to budget and time while at the same time achieving the required quality.

The practical effect of splitting the role however would appear to require the *Employer* to resolve any such conflicts given the direct reporting lines that both

---

[124] Clause 14.1.
[125] See Appendix 1 to Chapter 1 in Book 2.
[126] ECC3 clause 11.2(5).
[127] Refer to clause 42.1 and section 3.4.2 below.

the *Project Manager* and *Supervisor* enjoy. In fairness to the ECC, the Guidance Notes do concede, 'the roles of the *Project Manager* and the *Supervisor* may be combined where the objectives of the *Employer* are served by so doing'.[128]

It is not intended to go into great depth in this part of the chapter as to the role of the representatives in the context of 'control of quality' other than to list the actions,[129] which they are required to take, more of which is discussed in the relevant parts of this chapter.

3.4.1 Project Manager

The *Project Manager*'s actions with respect to quality include:

- giving instructions changing the Works Information,
- giving instructions resolving ambiguities or inconsistencies between the documents which are part of the contract,
- acceptance of the *Contractor*'s design for any parts of the *works* and for items of Equipment,
- acceptance of replacement key persons,
- acceptance of Subcontractors, the proposed subcontract conditions (in certain circumstances) and the proposed subcontract data (in certain circumstances),
- acceptance of method statements and the rest of the programme,
- giving instructions to stop or not to start any work,
- deciding the date of and certifying Completion,
- potentially, acceptance of the *Contractor*'s quality policy statement and quality plan if additional clauses are added to take this into account,
- potentially giving instructions to the *Contractor* to correct a failure to comply with the quality plan if additional clauses are added to take this into account,
- assessing amounts payable by the *Contractor* to the *Employer* in respect of costs incurred by the *Employer* and Others resulting from a test or inspection having to be repeated after a Defect is found,
- arranging for the *Employer* to give access to the *Contractor* to parts of the *works* already taken over by the *Employer* if needed to correct Defects,
- proposing changes to the Works Information so that Defects do not have to be corrected,
- arranging to have Defects corrected by people other than the *Contractor* where the latter has not corrected them within the *defect correction period*,
- requesting proof of the *Contractor*'s title to documents, Equipment, Plant and Materials prior to inclusion of the value of same in assessments of the amount due,
- giving instructions on how to deal with objects of value or of historical or other interest found within the site.

3.4.2 Supervisor

The *Supervisor*'s actions with respect to quality include:

- notifying the *Contractor* of any tests and inspections the *Project Manager* wishes to carry out,
- watching any tests done by the *Contractor*,
- notifying the results of tests and inspections,
- notifying the *Contractor* that Plant and Materials outside the Working Areas have passed any tests or inspections which the Works Information makes a precondition for their being brought to the Working Areas,
- giving instructions to search which may include:
  - uncovering, dismantling, recovering and re-erecting work,
  - doing tests and inspections which the Works Information does not require,
- notifying the *Contractor* of any Defects found,
- issuing the Defects Certificate,

---

[128]Page 7, project organisation, and paragraph 3.
[129]See Appendix 2 to Chapter 2 in Book 1.

> - marking documents, Equipment, Plant and Materials outside the Working Areas in order to secure for the *Employer* whatever title the *Contractor* has.

3.4.3 *Employer* The *Employer*'s actions with respect to quality include:

> - providing facilities, materials and samples for tests and inspections done by the *Supervisor* as stated in the Works Information,
> - giving access to the *Contractor* to parts of the *works* already taken over by the *Employer* if needed to correct Defects.

3.5 *Employer*'s supply

3.5.1 Plant and Materials, facilities and services

It is not unusual for *Employers* to provide things to the *Contractor*, either for inclusion in the *works* or to be used for (but not included in) the *works*. The terms used for these two categories of things by the ECC is 'Plant and Materials' and 'facilities and services', one definition of the latter being 'items provided by the *Employer* for use by the *Contractor* to Provide the Works but which the Works Information does not require the *Contractor* to include in the *works*', for example common-user construction plant and welfare facilities. Note that Equipment is provided by the *Contractor* by definition[130] and therefore items of plant provided by the **Employer** for use by the *Contractor* to Provide the Works but which the Works Information does not require the *Contractor* to include in the *works* cannot also be called Equipment.

The provision of Plant and Materials or facilities and services by the *Employer* brings with it attendant risks, which, like all risks, are best avoided unless the benefits justify such a course of action. The ECC recognises that there may be times when the *Employer* wishes to provide things to the *Contractor* but makes it clear where the risk lies in the event that things do not go according to plan. The two risks the *Employer* carries are:

(1)   that the *Employer* does not provide something which he is to provide by the date required by the Accepted Programme,[131]

(2)   an *Employer*'s risk event occurs (compensation event 60.1(14)) which in this context could be 'loss of or damage to [things] supplied to the *Contractor* by the *Employer* until the *Contractor* has received and accepted them'.[132]

The ECC sensibly leaves it to the Works Information to describe the 'things' to be provided by the *Employer* and a 'no earlier than' time for delivery. The *Project Manager* would then, as part of the tender assessment, need to check that the dates for requiring such things shown on the programme submitted by the tenderers with their tenders complied with the Works Information before affording them Accepted Programme status.

To protect himself from the second of the risks identified above, the *Employer* would be well advised to ensure that the Works Information includes, for things to be provided, clear procedures which leave the Parties in no doubt as to the point in time when the *Contractor* is deemed to have received and accepted the things (effectively the risk-transfer point in terms of responsibility for care).

3.5.2 Other contractors

The ECC defines 'Others' as 'people or organisations who are not the *Employer*, the *Project Manager*, the *Supervisor*, the *Adjudicator*, the *Contractor*, or any employee, Subcontractor or supplier of the *Contractor*'.[133] Consequently other contractors interfacing with the *Contractor* would be a sub-category of Others as defined.

Clause 25.1 of the ECC describes the extent of the *Contractor*'s obligations with regard to the interface with other contractors, namely:

---

[130]ECC2 clause 11.2(11); ECC3 clause 11.2(7).
[131]Compensation event in clause 60.1(3).
[132]Clause 80.1.
[133]ECC2 clause 11.2(2); ECC3 clause 11.2(10).

- to cooperate with other contractors in obtaining, providing and coordinating the information which he or they need in connection with his or their work,
- share the Working Areas with other contractors as stated in the Works Information.

Interface management is discussed further in Chapter 4 of Book 2.

To summarise, the *Contractor* is not responsible for the failure of other Parties to carry out their work in accordance with the Works Information unless the *Contractor*'s non cooperation causes the failure.

> The Works Information should include interface schedules for the work of other contractors.

3.6 Subcontracting

A Subcontractor is defined by the ECC[134] as:

ECC2 clause 11.2(9): 'A Subcontractor is a person or corporate body who has a contract with the *Contractor* to provide part of the *works* or to supply Plant and Materials which he has wholly or partly designed specifically for the *works*'.

In ECC the definition of a Subcontractor has been amended to:

'11.2.17 A Subcontractor is a person or organisation who has a contract with the *Contractor* to

- construct, or install part of the works
- provide a service necessary to Provide the Works or
- supply Plant and Materials which the person or organisation has wholly or partly specifically designed for the *works*.'

The change in wording clarifies that providing a service necessary to Provide the Works is included in the definition of a Subcontractor for the purpose of the contract.

This definition of a Subcontractor appears to embrace every person or corporate body with whom the *Contractor* has a contract in connection with the *works* **other** than suppliers of 'off-the-shelf' materials.

Given the tendency today for main contractors to subcontract a large part of the work, it is clearly important for the purposes of quality that care is exercised over the selection of Subcontractors; particularly since quality management systems and quality assurance may not be given sufficient emphasis in these organisations. It is not only the *Employer* who needs to be aware of the subcontracting arrangements. A *Contractor* will be fully liable to the *Employer* for the faults of a Subcontractor[135] and although he may in turn have a right to recover from the Subcontractor, if that Subcontractor has gone into liquidation or has insufficient assets, the *Contractor* will be left to bear the liability, however expertly drafted the terms of the subcontract.

It is therefore very much in a *Contractor*'s interest to minimise the likelihood of defects occurring and to be able to demonstrate clearly that he has fulfilled his contractual obligations. With these objectives in mind, the *Contractor* should not only implement a quality system covering his own activities, but also ensure either that his Subcontractors in turn implement their own quality system (preferable) or that his own quality system is sufficient to verify the Subcontractor's work. Clause 26.1 of the ECC states, 'if the *Contractor* subcontracts work, he is responsible for performing this contract as if he had not subcontracted'. This makes it clear that the *Contractor*'s quality management system, quality policy statement and quantity plan **must** cover any work, which the *Contractor* chooses to subcontract.

---

[134]ECC2 clause 11.2(9); ECC3 clause 11.2(17).
[135]Clause 26.1.

In general terms the ECC provides for the *Contractor* to subcontract all, some or none of the works provided the *Project Manager* accepts the proposed Subcontractors. Indeed, appointment of a Subcontractor for substantial work before acceptance by the *Project Manager* is expressly made grounds for termination of the contract.[136]

Traditionally, one way employers and their representatives have exercised a degree of control through contracts over the 'quality' of Subcontractors, particularly for specialist work, is by the process of nomination. The ECC however does **not** provide for the nomination of Subcontractors due, the ECC2 Guidance Notes explain, to 'the legal and practical problems of accountability which frequently ensue'[137] which in turn conflict with the principle (as embodied in clause 26.1) that the *Contractor* should be fully responsible for every aspect of the work he has contracted for. Alternatives under the ECC to nominating Subcontractors suggested by the ECC2 Guidance Notes include:

- leaving the *Contractor* freedom to subcontract as he thinks fit, with the *Project Manager* retaining some control over the identity of any proposed Subcontractors (see below),
- including lists of acceptable Subcontractors in the Works Information for particular parts of the works,
- providing for separate contracts with the *Employer* with the *Project Manager* managing the time and physical interfaces between them (not subcontracting within the accepted meaning of the term).

If a *Contractor*, under the ECC, decides to subcontract a part of the *works*, he is not permitted to appoint the proposed Subcontractor until the *Project Manager* has accepted both the Subcontractor (in all cases) and the proposed conditions of contract (unless a contract in the NEC family is to be used or the *Project Manager* has decided he does not require to see the proposed conditions). A reason for not accepting either the proposed Subcontractor or the proposed conditions of contract is that they will not allow the *Contractor* to Provide the Works in accordance with the Works Information.

It is open to some debate as to both why a *Contractor* would want to engage a Subcontractor who would not allow him to Provide the Works in accordance with the Works Information and on what basis the *Project Manager* could decide that a proposed Subcontractor would not allow the *Contractor* to Provide the Works. It is suggested that one way to reduce possible disputes on the issue is to include in the Works Information minimum requirements in respect of procurement procedures to be followed by the *Contractor* (to be developed by the *Contractor* as part of his quality plan) and minimum qualifying criteria for Subcontractors. An example is given below.

> The Subcontractor must have in place his own quality management system or ensure all employees are registered under construction skills certification schemes, or be able to demonstrate a minimum of five years experience in his chosen field on subcontracts of an equivalent magnitude.

> Include in the Works Information minimum requirements in respect of procurement procedures to be followed by the *Contractor*; for example, ensure value for money through term contracts or competitively tendering a minimum of three subcontractors.

---

[136]ECC2 clause 95.2(R13); ECC3 clause 91.2(R13).
[137]Page 4, some other changes, paragraphs 1 and 2.

3.7 Quality control

3.7.1 General Having defined, by way of the drawings and specification, the work to be undertaken and the standards of material and workmanship required, construction contracts have traditionally included provisions setting out:

- rights of access for the *Employer* and his representatives,
- powers and duties related to quality control,
- powers to deal with defective work,
- powers of enforcement and/or remedies in case the *Contractor* does not comply with his obligation to correct defective work.

The ECC includes such provisions which will be discussed under the headings of:

- access for the *Employer* and his representatives,
- tests and inspections,
- investigating defects and additional testing and
- quality procedures.

The provisions setting out the *Employer*'s powers to deal with defective work and of enforcement are dealt with in sections 3.8 and 3.10 respectively of this chapter.

3.7.2 Access for the *Employer* and his representative The ECC[138] provides very wide powers of access for the *Project Manager*, the *Supervisor* and others notified to the *Contractor* by the *Project Manager*, to work being done whether within the Working Areas or at the off-site premises of suppliers or Subcontractors. Such access is important for the purpose of checking on progress and for witnessing or carrying out tests and inspections.

3.7.3 Test and inspections The arrangements for inspection and testing should be clearly agreed between the Parties as part of the contract. The ECC seeks to achieve this by relying on the Works Information to state the following:

- The nature of the tests/inspections to be done.
- The timing of specified tests/inspections.
- Where the tests/inspections are to be done (e.g. within the Working Areas or before delivery to the Working Areas).
- Who is responsible for doing the tests/inspections (e.g. the *Contractor*, the *Supervisor*, an outside testing agency).
- Who is responsible for providing materials, facilities and samples for tests/inspections.
- The objectives of the tests/inspections, the testing procedures to be applied and the standards to be satisfied.

Technical specifications (i.e. the Works Information) should be prepared with a view to their function within a quality management system. They should therefore be practicable, realistic and capable of objective assessment by the *Contractor* without reference to the *Employer*'s representative (the *Supervisor*). This calls for great care in the drafting of the Works Information and the elimination of phrases such as 'to the satisfaction of' and 'in the opinion of' the *Employer*'s representative.

There is a very strong case to be made for the standardisation of the *Employer*'s technical specifications:

- to ensure that the information that the ECC requires to be included in the Works Information is actually present and
- to facilitate the development of quality system procedures and processes by contractors.

> Clause 40 deals only with tests and inspections that are required by the Works Information or the applicable law.

[138]In ECC2 clause 28.1; ECC3 clause 27.2.

It is important to realise that clause 40 in the ECC which deals with tests and inspections only applies to tests and inspections required by the Works Information or the applicable law (i.e. specified or statutory tests). It does not therefore apply to tests and inspections which the *Contractor* does at his own discretion or for his own purposes. This latter category of test/inspection could be quite extensive on contracts where the *Contractor* has a well-developed quality system in place where it is quite likely that the testing and inspection plan contained within the quality plan would require more extensive verification than that required by the Works Information and the law.

The *Project Manager*, of course, does have the power to request additional test and inspections to those required in the Works Information by issuing an instruction changing the Works Information[139] but such an instruction would be treated as a compensation event.[140]

In circumstances where the *Supervisor* is responsible for doing tests and inspections, clause 40.5 deals with the possibility that such tests or inspections cause unnecessary delay either to the work in hand or to any payment which is conditional upon a test or inspection being successful. Such an event becomes a compensation event if it can be established that the delay was 'unnecessary'.

Clause 40.3 of the ECC requires the *Contractor* and the *Project Manager* to notify each other before commencing tests and inspections and to notify each other of the results afterwards. This is in keeping with the contract's philosophy of ensuring that both Parties are kept informed of events and can respond quickly if, for example, tests reveal that any work does not comply with the standards specified in the Works Information.

Clause 41.1 of the ECC deals with tests and inspections which the Works Information expressly states are to be carried out on off-site Plant and Materials before the latter are delivered to the Working Areas. While this is a sensible provision designed to avoid the expense of transporting defective Plant and Materials back to the place of manufacture, the clause places the onus on the *Supervisor* to notify the *Contractor* that the particular Plant and Materials have passed the requisite tests and inspections. In instances where these tests and inspections are the responsibility of the *Contractor* this will require the latter to first notify the *Supervisor* of the results of the off-site tests and inspections but then to wait for the *Supervisor*'s acknowledgement (by way of notification) that the off-site Plant and Materials have indeed passed the prescribed tests and inspections. All a bit convoluted.

3.7.4 Notifying and investigating Defects and additional testing

It is very important under the ECC to understand what the contract means when it uses the term 'Defect'. Clause 11.2(15)[141] defines a Defect as:

- 'a part of the works which is not in accordance with the Works Information or
- a part of the works designed by the *Contractor* which is not in accordance with:
  - the applicable law or
  - the *Contractor*'s design which has been accepted by the *Project Manager*.'

> A Defect can only be determined with reference to the Works Information or the law.

There are a number of issues to appreciate from this definition:

- The quality standards stated in the Works Information (prepared by the *Employer*) provide the basis on which the existence of a Defect is judged. The importance of careful drafting of the technical standards is emphasised above.

---

[139] Clause 14.3.
[140] Clause 60.1(1).
[141] ECC3 clause 11.2(5).

- The definition **excludes** defects due to design for which the *Employer* is responsible. Such defects would have to be dealt with by way of instructions from the *Project Manager* and would invariably trigger the compensation event procedure.
- In instances where the *Contractor* is responsible for designing a part of the *works* and the *Project Manager* inadvertently accepts a design from the *Contractor* which does not comply with the Works Information in some respect, contractually a Defect will still arise if the *works* are constructed to the latter design. This is the effect of the 'two-limb' definition of a Defect and why it is vitally important that the *Contractor* ensures that the Works Information that he prepares and submits for acceptance complies with the Works Information prepared by the *Employer*.
- A Defect as defined may include a departure by the *Contractor* from procedures or instructions set down in the *Contractor*'s quality system (quality plan) for the *works*, where the *Contractor*'s quality system is included in the Works Information.

Having established what the contract means by the term Defect, how does it deal with the notification and investigation of Defects? Clause 42.2 states that 'until the *defects date* the *Supervisor* notifies the *Contractor* of each Defect which he finds and the *Contractor* notifies the *Supervisor* of each Defect which he finds'.[142] Simplicity itself, although somewhat unusually for construction contracts, this places a contractual obligation on the *Contractor* to admit openly that he has got something wrong. Although unusual for construction contracts this is entirely consistent with both the quality standards and the ECC philosophy of enabling problems to be identified as soon as possible in order that they can be dealt with properly.

> Both the *Supervisor* and the *Contractor* are obliged to notify each other of Defects.

If no Defect has been notified by the *Contractor* but the *Supervisor* suspects one may exist, clause 42.1 gives the *Supervisor* the power to instruct the *Contractor* to 'search' which may include:

- uncovering, dismantling, recovering and re-erecting work,
- providing facilities, materials and samples for tests and inspections to be done by the *Supervisor* (which by implication are additional to any stated in the Works Information),
- doing tests and inspections, which are additional to those which the Works Information requires the *Contractor* to do.

If a search is instructed and a Defect is discovered (i.e. non-compliance with the Works Information), the *Contractor* corrects the Defect and no compensation event arises. Conversely, if no Defect is discovered, the *Contractor* is entitled to a compensation event.[143]

3.7.5 Quality procedures Experience has shown that reliance on contracts alone as a means of assuring quality in construction has not met with the expectations of many clients of the industry. Consequently, forward-thinking clients have taken the lead in promoting the adoption, development and implementation of quality management systems by their suppliers (contractors/consultants). The quality management system and contractual approach to assuring quality are neither mutually exclusive of one another nor mutually dependent; they are concurrent means of assuring quality. However, an integrated approach is necessary to maximise the advantage and avoid the potential disadvantages from adopting both. This demands careful drafting of contracts and their supporting documentation (e.g. Works Information) by professionals who understand both the fundamentals of quality systems and contracts and the interaction between the two.

---

[142]In ECC3 the obligation for both parties changes to 'as soon as' in lieu of 'which' in ECC2.
[143]Clause 60.1(10).

3.8 Defective work

3.8.1 General Defective work will be discussed under the following headings:

- rejection,
- correction of Defects,
- defects liability period and
- concessions.

3.8.2 Rejection The ECC does not contain any express provision for rejection of a part of, or the whole of, the *works* in the event that the performance of the same is wholly unacceptable when judged against the requirements of the Works Information. Such an occurrence could arise when a *Contractor* responsible for designing and constructing the *works* fails to meet a performance specification expressed in terms of the efficiency of the completed *works* (not uncommon in the process and plant sector).

Instead the ECC includes for various possible remedies, in that Defects are addressed depending on the seriousness of the failure. These include:

- the *Contractor* is required to correct Defects (see section 3.8.3 below),
- the Defect is accepted and the contract price is reduced (see section 3.8.5 below),
- the *Contractor* is liable for the cost of having Defects corrected by Others (see section 3.10.4 below),
- low performance damages (see section 3.10.5 below).

3.8.3 Correction of Defects The *Contractor* has an obligation[144] to 'correct Defects whether or not the *Supervisor* notifies him of them'. This might seem like stating the obvious but it does emphasise the importance of the quality standards specified in the Works Information being capable of objective assessment, removing reliance on the *Employer*'s representatives as the arbiter of what is acceptable and what is not.

> The *Contractor* is required to correct Defects whether or not the *Supervisor* has notified him of them.

In terms of when the *Contractor* has to correct Defects, the ECC takes a practical approach, and **up to Completion** (of either a *section* or the whole of the *works*) generally leaves the timing of corrective works to suit the *Contractor*'s planning. There are however a number of incentives for the *Contractor* to correct Defects sooner rather than later. These are as follows:

- For the price-based main options of the ECC, only completed activities/ work which is 'without Defects which would either delay or be covered by immediately following work' is taken into consideration when assessing interim payments to the *Contractor*.[145]
- For cost-based options, Defects corrected after Completion are a Disallowed Cost.
- The *Project Manager* has no duty to certify Completion (of any *section* or the whole of the *works*) until the *Contractor* has corrected Defects, which would prevent the *Employer* from using the *works* (or Others from doing their work[146]). This has the double downside of both preventing the release of half of any retention held (where secondary Option P is used)[147] and increasing the likelihood of the *Contractor*'s exposure to delay damages (where secondary Option R is used).[148]

---

[144]Clause 43.1.
[145]ECC2 clauses A11.2(24) and B11.2(25); ECC3 clauses A11.2(27) and clause B11.2(28).
[146]ECC3 clause 11.2(2).
[147]ECC Option X16.
[148]ECC3 Option X7.

- The *Contractor* is required to show on each revised programme which he submits to the *Project Manager* for acceptance, how he plans to correct notified Defects.
- If there are many notified Defects, the *Contractor* might struggle to correct all of them within the *defect correction period* after Completion. It could be more efficient for the *Contractor* to correct the Defects at the time of notification while the resources are on Site and working in that area.

If a test or inspection shows that any work has a Defect, clause 40.4 states that the *Contractor* corrects the Defect and then repeats the test or inspection, which led to its discovery.

Completion of the *works* (or any *section* of it) is a significant milestone in determining the time within which the *Contractor* has to correct any Defects which still exist at that time or which subsequently come to light. This is because clause 43.1[149] places on the *Contractor* an obligation to correct notified Defects before the end of the '*defect correction period*'. The *defect correction period* begins at Completion for Defects notified before Completion (other than those which would prevent the *Employer* from using the *works* which by this time should no longer exist) and when the Defect is notified for other Defects – that is, those notified after Completion. The '*defect correction period*' being an italicised term in the ECC means that it must be given meaning by an entry in the Contract Data. Typically, *Employers* prescribe a period of two or three weeks.

Invariably there will be instances when Defects are discovered after Completion when the *Employer* has taken over the *works*. Clause 43.3[150] of the ECC deals with such a situation by requiring the *Project Manager* to arrange for the *Employer* 'to give access to and use of any part of the *works* which he has taken over if it is needed for correcting a Defect'.

> Defects notified before Completion must be corrected within the *defect correction period* after Completion.
>
> Defects notified before Completion that would prevent the *Employer* from using the *works* must be corrected before Completion.
>
> Defects notified after Completion must be corrected within the *defect correction period* after notification.

For maintenance contracts, where the more immediate correction of Defects is important, clause 43.1[151] can be amended so that Defects must be corrected within the *defect correction period* when notified, whether this is before or after Completion.[152]

3.8.4 Defects liability period   It should by now be realised that the term '*defect correction period*' as used by the ECC has a wholly different meaning from the identical phrase used in the ICE 6th Edition of Contract and from other such phrases as 'defects liability period' and 'maintenance periods' used in other standard forms. The *defect correction period* in the ECC is simply the time the *Contractor* has to correct notified Defects existing at or arising after Completion.

[149] ECC3 clause 43.2.
[150] ECC3 clause 43.4.
[151] ECC3 clause 43.2.
[152] ECC2 clause 43.1 states that Defects notified before Completion should be corrected after Completion. For some contracts, this is inadequate. For maintenance contracts in particular, it is important that Defects are corrected at the time of notification and not left until after Completion to be corrected. If it is important that Defects are corrected at the time of notification, clause 43.1 can be amended using Option Z: The third sentence in clause 43.1 is deleted and replaced as follows: 'This period begins when the Defect is notified.' The same amendment can be made to ECC3 clause 43.2 by deleting the second sentence of clause 43.2 and inserting 'This period begins when the Defect is notified.'

So how does the ECC define the period within which the *Contractor* is contractually liable to correct Defects arising in the *works*? It uses the expression '*defects date*' which, given its italicised status, requires it to be given effect by the *Employer* inserting a period (typically 52 weeks) in the Contract Data part one, a period which runs from the date of Completion decided by the *Project Manager*.

There is no obligation on the *Contractor* to notify Completion.

3.8.5 Concessions
Most traditional standard forms of contract do not cater adequately for rational decisions about defective work based on engineering considerations. Consider an example where contiguous piles forming the embedded walls to a shaft have been sunk outside the vertical tolerances required by the Works Information. Most engineers will appreciate that the cost, time and effort necessary to remedy this situation may be out of all proportion to the impact the Defect has on following works and indeed the completed *works*. Previously the *Employer*'s representative, faced with such a situation, had the following options:

- turn a blind eye,
- accept the Defect (although not permitted by the contract) with the consequent attendant implications if the result is unsatisfactory,
- play it 'by the book', demanding 'unreasonable' steps to correct the Defect out of all proportion to the likely impact of the Defect on the finished works. Such a course of action has the undesirable side-effect of encouraging the *Contractor* not to notify similar Defects in the future.

The ECC fortunately sweeps all this away by incorporating a very practical provision setting in train the possibility of acceptance of a Defect if the *Contractor* and the *Project Manager* so agree. Clause 44.1 simply permits either to propose to the other that the Works Information should be changed so that the Defect does not have to be corrected. The subtlety of this is that by changing the Works Information in such an instance, the Defect ceases to exist and consequently any later adverse implications of such 'acceptance' of the Defect are at the *Employer*'s risk. Obviously, great care is required on the part of the *Project Manager* who would invariably seek advice from his designers in such a situation. The machinery for such a concession is set out in clause 44.2 and involves the following steps:

(1)   The *Project Manager* and the *Contractor* decide whether they are prepared to consider a change to the Works Information so that a Defect does not have to be corrected (in practice the proposal is likely to be initiated by the *Contractor*).

(2)   If they are, the *Contractor* submits a quotation for **reduced** Prices (a saving to the *Employer*) or an earlier Completion Date or both to the *Project Manager* for acceptance (effectively the *Contractor*'s 'consideration' in return for the *Employer*'s loss of value).

(3)   The *Project Manager* either:
- accepts the quotation and instructs the necessary change to the Works Information, the Prices and the Completion Date accordingly or
- does not accept the quotation, leaving the *Contractor* either to correct the Defect or submit a revised (more favourable) quotation.

Any change to the Works Information is not a compensation event by definition (clause 60.1(1)).

The process has to be streamlined in order to reach a solution quickly before succeeding construction is superimposed on the 'defective' part.

3.9 Certification

The ECC does not contain terms such as 'practical completion', 'mechanical completion' or 'substantial completion'. Instead it places on the *Project Manager* an obligation to decide the date when the *Contractor* has:

- done all the work which the Works Information states he is to do by the Completion Date and

- corrected notified Defects, which would have prevented the *Employer* from using the *works* (and Others from doing their work).[153]

Having decided the date that both of the above states have been reached (the date of Completion), the *Project Manager* then has a duty to certify Completion within one week of that date.

It is hoped that by expressly stating in the Works Information the work the *Contractor* is to do by the Completion Date, the uncertainty associated with such terms as 'substantial' and 'practical' completion will be avoided. The state of Completion clearly has not been reached if there remain uncorrected, notified Defects which would prevent the *Employer* from using the *works*.

ECC3 adds a default: if the work which the *Contractor* is to do by the Completion Date is not stated in the Works Information, Completion is when the *Contractor* has done all the work necessary for the *Employer* to use the *works* and Others to do their work.[154]

Clause 11.2(16)[155] of the ECC defines what is meant by a Defects Certificate, which is either:

- a list of Defects that the *Supervisor* has notified before the *defects date* which the *Contractor* has **not** corrected or
- if there are no such Defects, a statement that there are none.

At first sight it might appear that the Defects Certificate is the equivalent of the defects correction certificate under the ICE 6th Edition conditional contract or the certificate of completion of making good defects under JCT Forms. It is not! Those certificates are only issued when the *Contractor* has fulfilled his obligations to make good Defects, whereas under the ECC the Defects Certificate is issued on a set date as a **record** of whether or not the *Contractor* has fulfilled his obligations. If the *Contractor* has fulfilled his obligations the Defects Certificate will simply contain a statement that there are no Defects to be corrected. If the *Contractor* has failed to fulfil his obligations then the Defects Certificate will list the particular Defects which the *Contractor* has failed to correct. The latter gives rise to the right for the *Project Manager* to have the uncorrected Defects corrected by others (see section 3.10.4 below).

The timing of the issue of the Defects Certificate by the *Supervisor* is stated in clause 43.2[156] as the later of:

- the *defects date* or
- the end of the last *defect correction period* (which may be later than the *defects date* if it commenced just before the *defects date* and therefore ends after it).

> In ECC the words:
>
> 'The *Employer*'s rights in respect of a Defect which the *Supervisor* has not found or notified are not affected by the issue of Defects Certificate' have been added to clarify that no transfer of liability occurs through the issue of the Defects Certificate.

The purpose then of the Defects Certificate is to put on record the state of the *works* at the date at which the *Contractor*'s entitlement to correct Defects expires. It is not therefore to be taken as a certificate of confirmation of fulfilment of the *Contractor*'s obligations. However, the effect of the Defects Certificate is similar to that in other contracts, in that it triggers the release of the

---

[153] ECC3 clause 11.2(2).
[154] ECC3 clause 11.2(2).
[155] ECC3 clause 11.2(6).
[156] ECC3 clause 43.3. This clause has the added words: 'The *Employer*'s rights in respect of a Defect which the *Supervisor* has not found or notified are not affected by the issue of Defects Certificate.'

final part of retention money (where secondary Option P is used)[157] and sets the date for the expiry of various other obligations.

It is appropriate at this time to consider the subject of latent defects – that is, defects that only appear after the Defects Certificate has been issued. In common with other standard forms of contract, the ECC does not expressly exclude the *Contractor*'s liability for latent defects and consequently the *Contractor*'s liability follows the law applicable to the contract (subject to any limitation arising from clause 21.5[158] which only appears to apply to Defects due to the *Contractor*'s design – not workmanship).

3.10 Enforcement

3.10.1 General

The fundamental significance of contracts and the law in general is enforceability by the courts. The term 'enforceability' as used in this context however is somewhat misleading in that it does not mean that the contract, or indeed ultimately the courts, will actually ensure that an agreement is fulfilled as intended. For example, a contract or the courts will not ensure that a contract to construct particular *works* will actually result in those works being constructed in accordance with the specified requirements. Enforceability, from the viewpoint of the *Employer*, means that in the event of the *works* not being completed in accordance with the specified requirements, the contract will provide certain remedies or the courts will award financial compensation in respect of any loss or damage suffered by the *Employer*. It is left to the disappointed *Employer* to actually deal with the defects.

This final part of this chapter therefore deals with the contractual incentives and the remedies available to the *Employer* when things do not go to plan and it becomes necessary to take 'enforcing' action. These remedies will be dealt with in an order starting with what might be considered the 'least serious' and working up to 'most serious'. They will be considered under the following headings:

- incentivisation through certification,
- removal of employees,
- correction of Defects by Others,
- low-performance damages,
- termination of the *Contractor*'s employment.

3.10.2 Incentivisation through certification

Although it has already been noted that the timing of the correction of pre-Completion Defects is largely a matter for the *Contractor*, there are powerful incentives introduced through the certification process (both of interim amounts due and of Completion) that make it in the *Contractor*'s financial and commercial interest to correct Defects sooner rather than later (refer to section 3.8.3 above).

3.10.3 Removal of employees

Since most quality problems can be traced back to people, be it due to the preparation of inadequate quality procedures or a failure to ensure that procedures are adhered to, it is reassuring to see that the ECC gives the *Project Manager* the power to instruct the *Contractor* to remove any employee, having stated his reasons for doing so (refer to clause 24.2). If there was any doubt over whether the *Project Manager*'s powers extended to include employees of Subcontractors, this is dispelled by the second sentence of clause 26.1, which states that 'this contract applies as if a Subcontractor's employees and equipment were the *Contractor*'s'. The *Contractor* is obliged to ensure within 24 hours of the *Project Manager*'s instruction that the employee in question has no further connection with the contract. The *Contractor* may also be required to remove the employee immediately.

3.10.4 Correction of Defects by Others

Clause 45.1 of the ECC is similar to the type of clause found in most construction contracts, entitling the *Employer* to recover the cost of making good Defects

---

[157] ECC3 Option X16.
[158] ECC3 clause 21.5 is deleted and has been replaced by Option X18.3.

if the *Contractor* has failed to correct them within a prescribed time, in this case the *defect correction period*. Since in the majority of cases uncorrected Defects will be a post-Completion issue, it will fall to the *Project Manager* to offset the cost of having the uncorrected Defects corrected by other people, against the release of the second half of the retention money assuming ECC2 Option P[159] is included in the contract. The *retention percentage* should consequently be sufficient to produce an appropriate fund, which remains in the *Employer*'s hands until the *defects date* when the extent of the cost of uncorrected Defects is known.

Retention bonds, which are gaining in popularity with some Employers as an alternative to the traditional retention arrangements, are not directly catered for by the ECC although they could be added through Option Z and by including the required form in the Works Information.

3.10.5 Low performance damages

Under the ECC the possible *Employer*'s remedies for low performance are as follows:

- The *Contractor* is required to correct Defects – clause 43.1 (refer to section 3.8.3 above).
- The contract price is reduced following the 'acceptance' by the *Project Manager* of a Defect which is not corrected – clause 44.[160]
- The *Contractor* is liable for the cost of having Defects corrected by others – clause 45.1 (refer to section 3.10.4 above).
- Low performance damages – secondary Option S.[161]

The low performance damages secondary Option is most likely to be used in circumstances where the *Contractor* has full design responsibility and the *Employer* has expressed his requirements for the completed works by way of performance criteria included in the Works Information. They feature most commonly in process and plant contracts but are occasionally found in construction contracts. Option S[162] of the ECC states that 'if a Defect included in the Defects Certificate shows low performance with respect to a performance level stated in the Contract Data [part one], the *Contractor* pays the amount of low performance damages stated in the Contract Data [part one]'. The consequence of this is that where the performance of the *works* in use fails to reach a specified level due to a design or other fault of the *Contractor* and the Defect is not corrected (i.e. it is listed in the Defects Certificate), the *Employer* should be able to recover the damages he suffers in consequence, a genuine pre-estimate of which should be included by the *Employer* in the Contract Data part one.

Any deduction of low performance damages is made in the assessment of the amount due following the issue of the Defects Certificate which again emphasises the need to consider making provision for the existence of an adequate retention fund at this time from which to set off any damages due.

3.10.6 Termination of the Contractor's employment

Although this would be a last resort, the ECC does provide for the Contract to be terminated if, subject to a four-week period to rectify a particular default, the *Contractor* has 'substantially failed to comply with his obligations' (clause 95.2 (R11)).[163] Although the language used is fairly general, it is submitted that persistent failure by the *Contractor* to comply with the accepted quality plan, particularly where in reliance on such compliance the *Employer* has reduced the levels of external supervision, would amount to a substantial failure by the *Contractor* to comply with his obligations. It would of course be helpful to the *Employer*'s case if he could demonstrate evidence of an intolerable level of Defects resulting from the *Contractor*'s failure to comply with the quality plan.

[159] ECC3 Option X16.
[160] Refer to section 3.8.5 above.
[161] ECC3 Option X17.
[162] ECC3 Option X17.
[163] ECC3 clause 91.2(R11).

3.11 NEC 3rd Edition

The following table indicates the changes made in ECC in relation to quality control.

NEC2 clause	NEC3	Comments
40.1	Wording amended as follows: 'The subclauses in this clause only apply to tests and inspections required by the Works Information or the applicable law.'	The rewording clarifies the intent of the clause.
42.1	Wording amended to make it clear that this clause: • applies until the *defects date* • relates to a search for a Defect.	
42.2	The words 'as soon as' have been added to the clause to make it an obligation to take action immediately and not to notify at a later point in time.	The inclusion of the words 'as soon as' re-emphasise the theme of the NEC which is all about communication between the parties and the taking of action at the most appropriate time.
43.1	The first sentence of the ECC2 clause remains unchanged. This makes an obligation for the Contractor to correct a Defect. The latter part of the original clause is now included in a revised clause 43.2.	First part of the clause remains the same with the latter part now moved to 43.2.
43.1	New 43.2 ECC2 wording deleted *in toto* and replaced with the latter part of what was clause 43.1.	Last two sentences of clause 43.1 in ECC2.
43.2	Renumbered to 43.3. New last sentence added which reads: 'The *Employer*'s rights in respect of a Defect which the *Supervisor* has not found or notified are not affected by the issue of the Defects Certificate.'	Reinforces the point that the issue of the Defects Certificate does not relieve the *Contractor* of his contractual obligation to Provide the Works in accordance with the Works Information.
43.3	Renumbered to 43.4 with changes to the wording.	The intent remains fundamentally the same. The emphasis is now on the *Employer* to 'allow' rather than 'give' and for the *Employer* to allow access to 'a' rather than 'any' part of the *works*, if 'they are' needed for correcting a Defect. The last part of the clause has now been amended to read: 'In this case the *defect correction period* begins when the necessary access and use have been provided.' This wording replaces the obligation for the *Project Manager* to extend the period for correcting a Defect.
45.1	The wording has been amended.	The wording of this clause has be changed so that if the *Contractor* has been given access to correct a Defect and has not done so within the *defect correction period* then the *Project Manager* assesses the cost to the '*Employer*' of having the Defect corrected by other people and the *Contractor* pays this amount. It also clarifies that the Works Information is treated as having been changed to accept the Defect.
	New clause 45.2: 'If the *Contractor* is not given access to correct a notified Defect before the *defects date*, the *Project Manager* assesses the cost to the *Contractor* of correcting the Defect and the *Contractor* pays this amount. The Works Information is treated as having been changed to accept the Defect.'	This clause gives the *Employer* the right to deduct the costs of Defects not corrected because the *Employer* did not or could not give the *Contractor* access to correct notified Defects before the *defects date*. Some *Employers* have had difficulties in allowing the *Contractor* access to a part of the *works* after Completion due to security or operational issues or the like. It should also be noted that it is the *Project Manager* who assess the costs to the *Contractor*.

NEC2 clause	NEC3	Comments
Other related changes		
18.1 Health and safety	Clause deleted *in toto* and replaced as clause 27.4.	There are no implications to the meaning of the clause now that it is placed in a different section of the core clauses.
31.2	The second bullet point of ECC2 which reads: • 'for each operation, a method statement which identifies the Equipment and other resources which the *Contractor* plans to use' has been deleted and replaced with: • 'for each operation a statement of how the *Contractor* plans to do the work identifying the principal Equipment and other resources which he plans to use.'	Although not specifically called a method statement, the requirement in ECC3 is no different from the requirement in ECC2.

4 Disputes and dispute resolution

Synopsis

This chapter:

- Emphasises the importance of early dispute resolution to the successful outcome of a contract

- Considers the common sources of dispute

- Considers how the ECC has been designed to reduce the incidence of disputes

- Examines how the ECC provides for the resolution of disputes

- Looks at the implications for the dispute resolution process as a result of the new Housing Grants, Construction and Regeneration Act 1996

- Looks at ECC changes in relation to adjudication

4.1 Introduction

In recent years the construction industry has built up a reputation for being adversarial. The industry has spent millions upon millions of pounds in disputes that add no value to the construction process.

Most disputes at site level are founded either on:

- differences in interpretation of the documents forming the contract,[164]
- differences of opinion over the financial and/or time effects of supervening events.[165]

Such disputes divert considerable resources, sometimes at the expense of the ongoing construction. They cause budgetary uncertainty for employers and financial difficulties to contractors (and their subcontractors). Consequently, one of the few things that most people in the construction industry do agree on is that disputes are not a good thing. Indeed, most initiatives over recent years aimed at improving the performance of the construction industry have concluded that the adversarial culture has to change.

It should therefore come as no surprise that one of the fundamental objectives of the NEC's Engineering and Construction Contract was that its use should minimise the incidence of disputes and therefore improve the certainty about the outcome of the contract for both parties. However, the ECC sensibly recognises that as long as two people could place different interpretations on or have different opinions on the same issue then disputes would continue to arise. At the same time, given the potential damage such disputes present to the collaborative working principles upon which the ECC is founded, the drafting team realised that disputes, having arisen, need to be resolved quickly and by a process which in principle is accepted as fair by both parties.

> Fundamental objective of the ECC is that its use should:
> - minimise the incidence of disputes and
> - improve certainty of outcome for both Parties.

4.2 How disputes arise

4.2.1 Introduction

Two of the main seedbeds of disputes under an ECC contract are the Works Information and the Site Information.

It is vitally important that the information provided to the *Contractor* at tender stage is as complete as it can possibly be. This means that a great deal of effort needs to be put into the preparation of information which will form part of the contract documents.

These requirements lead to a need for openness in the preparation of documentation. If the *Employer* genuinely does not know about something, he should state so and any later addition can be managed through the compensation event procedure. Alternatively, the *Employer* could give assumptions upon which to base the bid. A *Contractor* is in no better position to manage the 'risk' than the *Employer* is so there is no point in 'passing it on to him', because it will eventually come back to the *Employer* to manage.

This philosophy is at odds with traditional professional training which encouraged the use of 'all-embracing' preambles to contract documents to cover everything stated and not stated.

Another main area where risk arises is in the administration of the contract. An action/inaction under the contract can be grounds for a dispute to arise.

---

[164]See section 4.2.2 below.
[165]See section 4.2.3 below.

> Ensure that information provided to the *Contractor* at tender stage is as complete as possible.
>
> Good documentation and good administration are essential in avoiding disputes.

The majority of disputes are probably avoidable but before we consider how to avoid them, we need to understand better how they arise.

4.2.2 Interpretation of documents

The first source of disputes mentioned earlier is those that arise from a difference in interpretation of the documents forming the contract.

The contract is drawn up to define what is required to be done in return for what payment; that is, the duties and responsibilities to be undertaken by each Party and (to some extent) what is to happen should they fail to exercise them. Where the definition of what is to be done is incomplete, the contract gives the *Employer*'s representative certain powers to supply further information and also to vary the work to be done. Risks, which may be encountered in the execution of the *works*, are allocated (both contractually and financially) between the Parties. The physical context within which the *works* are to be carried out and any constraints on the work to be done should also be stated.

Generally it is the words used that matter – not those that could have been or even should have been used. The courts are inclined to take the words used at their face value, to assume that if used, they were intended, that the same word has the same meaning throughout and that if different words are used, then different things are meant. They work on the basis that the words of the contract were agreed between both Parties to the contract and that being so it is not open to either Party subsequently to complain that the responsibilities imposed are onerous.

It is convenient to think that the above applies only to the *conditions of contract* but of course it applies to **all** the documents forming part of the contract. In the case of the ECC this means the documents stated as being part of the contract, for example:

- The Form of Contract/Articles of Agreement.
- The core clauses and the main and secondary Option clauses of the ECC.
- The Contract Data part one and all documents referred to therein, principally the Works Information and the Site Information.
- The Contract Data part two and all documents referred to therein, including any Works Information for the *Contractor*'s design, the first programme submitted for acceptance and the pricing document in the form of an *activity schedule* or *bill of quantities*.

Ironically, it is often the *conditions of contract* that prove to be the least fertile seedbed for disputes, probably explained by its careful legal drafting and (at least in the case of standard conditions) the Parties' familiarity with their meaning, their responsibilities under them and the recognised allocation of risk. The more fertile sources of dispute tend to be those documents that have to be prepared each time to suit the specific requirements of the individual contracts. With the ECC this means the Works Information, the Site Information and to a lesser degree the Contract Data. It is worth considering the definition given to these supporting documents by the ECC.[166]

> The interpretation of documents applies to **ALL the documents stated as being part of the contract not just the** *conditions of contract*.

---

[166] As defined in ECC2 clauses 11.2(5) and 11.2(6); ECC3 clauses 11.2(19) and 11.2(16).

4.2.2.1 Works Information

'Works Information is information which either specifies and describes the *works* or states any constraints on how the *Contractor* provides the works.'[167]

Works Information will therefore typically include the general specification, the preliminaries, the materials and workmanship specification, the drawings, any *Employer*'s requirements in respect of parts of the *works* which the *Contractor* is to design and any other document which describes what the *Contractor* is to construct and which describes any constraints on how he is to go about it.

Most disputes involving the Works Information are therefore rooted in one of the following:

- The Works Information is deficient in some respect, for example unclear as to who is responsible for certain actions (e.g. obtaining consents) or unclear as to the detail or the quality standards to be achieved.
- There exists ambiguity or inconsistency in or between the documents that comprise the Works Information.
- The *Contractor* contends that the Works Information requires him to do something that is illegal or impossible.

It is probably fair to say that in pursuit of the 'flexibility' objective, the ECC places greater reliance on the Works Information as a source of supplementary information than more traditional contracts do on its equivalent. For example, details of testing are not included in the *conditions of contract*, but must be drafted by the *Employer* and included in the Works Information. It therefore follows that greater skill and care is required in the drafting of the Works Information (in all its guises) if certain provisions of the ECC are to be effective. Some examples of the importance the ECC attaches to the Works Information are given in the table below.

Clause No.	Comments
11.2(15)[168]	A part of the *works* not in accordance with the Works Information is a Defect. It follows that the quality standards set out in the Works Information provide the basis on which the existence of a Defect is judged. Problems arise when the Works Information is silent on quality standards.
11.2(13)[169]	Completion is when the *Contractor* has done all the work which the Works Information states is to be done before the contractual Completion Date.
18.1[170]	The *Contractor* is to act in accordance with the contract specific health and safety requirements stated in the Works Information.
21.1	The *Contractor* is to design such parts of the work as stated in the Works Information.
25.1	The *Contractor* is to share the Working Areas with Others as stated in the Works Information.
40.2	The *Contractor* and the *Employer* provide materials, facilities and samples for tests and inspections as stated in the Works Information.

> The interaction between the *conditions of contract* and the Works Information means that disputes can occur when the Works Information does not contain the information it should to give effect to the *conditions of contract*.

[167]ECC2 clause 11.2(5); ECC3 clause 11.2(19).
[168]ECC3 clause 11.2(5).
[169]ECC3 clause 11.2(2).
[170]ECC3 – this clause has been renumbered to clause 27.4.

The quality of the drafting of the Work Information is very important. Some common problems include the following:

- A lack of precision as to what is required, a common example being quality standards that are described subjectively in terms such as 'to the satisfaction of the *Engineer/Architect*'.
- An inconsistency in style between the various parts of the Works Information as a result of different people's contribution. There must be an overall editor/coordinator of the contract documents during their preparation if a 'patchwork quilt' effect is not to result.
- A total ignorance of the fact that the Works Information should be compatible with the *conditions of contract* and consequently the use of different terms and expressions that conflict with the responsibilities and risk allocation set out in the conditions.
- A simple failure to appreciate how much information the *Contractor*'s estimator needs in order to prepare a reliable tender. It must be realised that another party is committing himself to translate the documents into physical reality for a price.

It has been said that the successful outcome of a contract is largely dependent on the managerial effort applied to the pre-planning and preparation of the contract documentation. Prevention of disputes is after all more economical than having to resolve them later. Given this fact, it is still incredible how poorly planned the pre-contract phase of most projects remains, with the preparation of tender documentation still viewed by many *Employers* and *Project Managers* as a simple activity of short duration squeezed in between the completion of design and the commencement of construction.

Having considered disputes rooted in the Works Information let us now consider the other information traditionally prepared by the *Employer* and unique to each contract, namely the Site Information.

> The ECC places great emphasis on the Works Information as a source of supplementary information, far more so than traditional contracts.

4.2.2.2 Site Information   The ECC definition of the Site Information is simply information that describes the Site and its surroundings, the Site being that area within the *boundaries of the site* identified in the Contract Data. Consequently, the Site Information would include details of such matters as:

- any existing buildings/other structures at the Site,
- any existing buried or above ground services,
- soil characteristics,
- the levels of interfaces between different geological strata,
- groundwater levels,
- the presence of any bodies of water within or surrounding the Site and seasonal water levels,
- interpretative soil investigation reports.

It is easy to fall into the trap of thinking that Site Information relates to things (natural and man-made) that are pre-existing at the Site before any work is commenced. On a multi-contract project, details of the works constructed by the previous contractor become the Site Information for the follow-on contractor. Consequently, on a project for a new below-ground transportation system, the layout of the tunnels and the details of the tunnel linings, all constructed by the preceding 'civils' contractor, become the Site Information for the succeeding contractor engaged to install the mechanical and electrical system. For this reason, as well as to meet the requirements of the CDM Regulations, it is important to keep good records of the works actually being constructed by the different contractors on multi-contract projects.

On the basis that the Site Information should represent a factual account of the Site and its surroundings, it is surprising how many disputes arise from the encountering of physical conditions which to use the common expression 'could not reasonably have been foreseen by an experienced contractor'. So why should this be?

Historically, employers have invested too little money at the front end of projects which is where any useful site investigation work is of most use, thereby ignoring the old maxim that 'money spent earlier buys more than money spent later'. In addition to the problem of not allocating sufficient monies to site investigation, is the related problem of collecting information that is largely irrelevant to the construction of the works. A common problem is for the site investigation to focus on design of the permanent works with little or no thought as to what information would be useful to the contractor in order to determine the most economic temporary works solution and working methods.

To emphasise the importance of comprehensive and relevant Site Information to the avoidance of disputes, consider the 'physical conditions risk' carried by the *Contractor* under an ECC. Putting weather conditions to one side (since these are dealt with separately), if the *Contractor* is to notify successfully for more money and/or time, he has to be able to persuade the *Project Manager* that the physical conditions actually encountered within the Site are so different that an experienced *Contractor* would have judged them to have such a small chance of occurring that it would have been unreasonable for him to have allowed for them. Recognising that this still leaves room for some interpretation, the ECC seeks to narrow the boundaries surrounding this provision by stating that for the purposes of assessing a compensation event:

> 'In judging the physical conditions, the *Contractor* is assumed to have taken into account
>
> - the Site Information
> - publicly available information referred to in the Site Information
> - information obtainable from a visual inspection of the Site and
> - other information, which an experienced contractor could reasonably be expected to have or to obtain.'[171]

Ignoring the obvious and the catch-all (third and fourth bullet points respectively) it should be clear that the test of what should have been foreseeable by the experienced *Contractor* still relies heavily on information in or referred to in the Site Information and the advice remains for *Employers* to buy the most comprehensive and relevant site investigation appropriate to the circumstances. However, to state a sum of money that should be spent on site investigation as a percentage of the value of the overall work can be misleading. It is far better to approach the subject from the risk analysis perspective.

> Take a new motorway contract where the designers are seeking to achieve a new vertical alignment that ensures a cut-fill balance; that is, using the excavated material derived from the cuttings as fill in the new embankment with a minimum of material to be disposed of off-site.
> It is clearly important to the *Employer*'s budget and programme for the contract to have a high level of confidence that sufficient quantities of acceptable material are present in the proposed cuttings. The consequence of this not being the case is the high cost of the disposal of quantities of unacceptable materials off-site (landfill tax included) and the additional cost of 'importing' acceptable fill materials to the site. Given the 'risk exposure' in this instance, it might well prove desirable to invest in a thorough site investigation and if necessary alter the vertical alignment to achieve the earthworks balance objective. Comprehensive site investigation is invariably a sound investment since if it narrows the definition of the likely conditions to be encountered, it should accordingly reduce the amount of risk monies included in the *Contractor*'s tender for what could otherwise be perceived as widely varying conditions.

[171]Clause 60.2.

So what are the common sources of dispute associated with the Site Information?

- Where the Site Information does not accurately represent the actual physical conditions encountered. As noted, very often this is a result of an inadequate/irrelevant site investigation, but now and again it is just the result of something unexpected. This is more common with below-ground civil engineering works, where no matter how thorough the site investigation, until the *works* are actually executed, the true nature of the conditions will never be known for certain. It was this feature of civil engineering works that has traditionally led to contracts for such work being accompanied by a bill of quantities containing provisional quantities of the work, all subject to admeasurement.
- Inconsistency or ambiguity in or between the documents, which form part of the Site Information.

It should be noted that most disputes surrounding the Site Information are not black-and-white cases. It is for this reason that the ECC2 Guidance Notes introduce the concept of 'boundary limits'. These should be introduced to the contract through the use of secondary Option Z additional conditions.

> Good early and comprehensive site investigation for the permanent and temporary works is essential.

4.2.2.3 Contract Data The final document mentioned earlier that is unique in its content to each project is the Contract Data. This comes in two parts, part one prepared by the *Employer* and sent out with the invitation to tender letter and accompanying tender documentation and part two prepared by the *Contractor* and submitted with the tender submission. The Contract Data is not defined in the ECC conditions, but this belies its importance. Its purpose is to provide key information as required by the *conditions of contract* and which is specific to a particular contract. It is absolutely key to the effective operation of the ECC that the terms in italics contained in the *conditions of contract* are given their meaning by the related entry in the Contract Data. For example, the following are addressed in the Contract Data:

- the main Option and secondary Options applicable to the particular contract,
- the names of the *Employer, Contractor, Project Manager, Supervisor* and *Adjudicator,*
- where to find the documents comprising the Works Information and the Site Information,
- the *starting date, possession dates*[172] and *completion date(s),*
- the *method of measurement* (if an ECC main Option utilising a *bill of quantities* is required),
- the amount of delay damages (Option R)[173] payable by the *Contractor* if the *works* are late and the amount of low performance damages (Option S)[174] payable if the *works* do not meet stated performance levels,
- the names of the *Contractor*'s key people,
- any Works Information in respect of designs for which the *Contractor* is responsible,
- the identity of the *activity schedule* or *bill of quantities* as appropriate together with the tendered total of the Prices,
- the identity of the first programme to be submitted.

The ECC sensibly includes a pro-forma Contract Data and consequently the likelihood of getting it wrong should be small, most errors arising from a misunderstanding of the information to be inserted. Getting it wrong is serious, however, since unlike the Works Information, which can be changed by an instruction given by the *Project Manager*, once the contract has been let, the

---

[172]ECC2 called *access dates.*
[173]ECC3 secondary Option clause X7.
[174]ECC3 secondary Option clause X17.

Contract Data can only be changed by agreement between the *Employer* and the *Contractor*.

One common mistake made by drafters of the documents comprising an ECC contract is to mix up information between the Site Information and the Works Information, for example giving information describing the Site and its surroundings in documents identified by the Contact Data as being Works Information (and vice versa). This is something that must be avoided.[175]

> The *Employer* might have commissioned the most comprehensive and relevant site investigations possible but if he then includes the findings in a document referred to as Works Information, he will not be able to rely on it as Site Information when seeking to counter a compensation event notification from the *Contractor* contending changed physical conditions. This should be a very sobering thought!

It is essential to ensure that the data in Contract Data part one and Contract Data part two is complete. It is not uncommon to find that some data are not inserted by the *Contractor* at the time of tender, for example components of cost for the Shorter Schedule of Cost Components.

> Ensure the Contract Data has been fully completed at tender and tender assessment stages. Works Information and Site Information should be kept separate. Failure to do so might have implications on how a compensation event may be assessed.

4.2.3 Cost and time effect of disputes

The second source of dispute identified is those arising from differences of opinion over the financial and/or time effects of events that arise once the contract has been let. In this instance, since the hurdle as to whether entitlement exists contractually for any particular case has been overcome, it remains only to establish its financial and time effects.

Traditionally, price-based contracts have sought to assess the financial effects of variations and other events at the *Employer*'s risk as follows:

- If the nature of the works affected by the 'event' and the conditions under which it is required to be undertaken are unchanged from those pertaining when the *Contractor* prepared his tender then such rates would be used to value the effects of the variation or change.
- If one or other of the nature or the conditions is dissimilar then the contract rates and prices are used as the basis for assessing the value of the variation or change.
- If neither the nature of the work nor the conditions are similar then the Engineer/Architect would be responsible for ascertaining a fair evaluation, but in doing so would seek to ensure that as far as possible the valuation was still related to the *Contractor*'s original contract rates and prices.
- As a last resort in circumstances where none of the above were feasible, the variation or change would be valued on a cost-plus or 'day work' basis.

This process resulted in many disputes over the applicability of the contract rates, usually contained in a bill of quantities, a document based on the misapprehension that all a *Contractor*'s costs are proportional to the quantities of the various elements of the work.

A *Contractor* losing money or seeking to earn inflated profits could always be relied upon to come up with all sorts of plausible reasons why the contract rates were not applicable to the varied work. The *Employer*'s representative often found it difficult to counter such arguments given the veil that the bill of quantities throws across the *Contractor*'s true costs and the manner in which they are incurred.

---

[175]See Chapter 4 of Book 2, Works Information guidelines.

If agreeing the financial effects of change was a challenge, then agreement of the time effects was nigh on impossible given the traditional scant regard paid to monitoring progress against an original and meaningful programme and it being in the *Contractor*'s interest to address the delaying effects for which the *Employer* was responsible later rather than sooner, once delays caused by his own inefficiencies were less prominent in the memory of the *Employer*'s representative.

It is perhaps sad that until very recently far more space on construction industry bookshop shelves and far more seminar time were devoted to the subject of 'claimsmanship' than to improving the performance of the construction industry for the benefit of its customers. Although it looks like this trend has been reversed, disputes over both money and time still comprise a large proportion of the total of all disputes at site level.

4.3 How the ECC seeks to reduce the incidence of disputes

When designing the ECC the drafters intended it to be flexible and clear and to promote good management. One example of this is the avoidance in the contract of such phrases as 'in the opinion of the Engineer'. Instead the duties of the *Project Manager* are clearly set out and the criteria on which his decisions are to be based are stated specifically, not left to a general concept of acceptability.

In many instances, the ECC will serve to reduce the incidence of disputes by virtue of its two founding principles, both of which have a major impact upon the objectives of stimulating good management. These principles are:

(1) foresight applied collaboratively mitigates problems and shrinks risk and

(2) clear division of function and responsibility helps accountability and motivates people to play their part.

Some practical examples of how these principles serve to reduce the incidence of disputes follow.

4.3.1 Early warning The early warning provision of the ECC places an obligation on both the *Project Manager* and the *Contractor* to give to the other a notification of any matter which could increase the price the *Employer* pays, delay Completion or impact on the finished quality of the *works*.[176] This is intended to be a practical device to stimulate early joint consideration of unforeseen problems.

Joint consideration of the problem should lead to joint agreement as to the best solution and consequently the necessary action to avoid the problem or reduce its impact. In addition to shrinking the risk to the price, the programme and the quality of the *works*, the early warning provisions, by involving the *Project Manager* in the decision making, reduce the possibility of him deciding at a later date with the wisdom of hindsight that the *Contractor* did not deal with the problem in the most cost- and/or time-effective manner.

4.3.2 Valuing changes Many disputes arise over the assessment of the financial and time effects of variations and other changes at the *Employer*'s risk under the contract. These disputes invariably centre on the applicability of the contract rates to the changed situation and the ineffective use of a programme to monitor progress and plan the future work.

Under the ECC, the traditional basis of valuing variations using tendered bill of quantities rates is discarded in favour of valuation according to the full effect of the variation on timing and methods of work, and the use of resources.[177]

---

[176] ECC3 – a new fourth bullet point and additional words have been added to clause 16.1 as follows: 'delay meeting a Key Date or the *Contractor* may give an early warning by notifying the *Project Manager* of any other matter which could increase his total cost. Early warning of a matter for which a compensation event has previously been notified is not required'.

[177] Although rates can be used by agreement in the assessment of compensation events (ECC2 clause B63.9 and ECC3 clause B63.13).

Such a means of valuation relies on the *Contractor* maintaining a comprehensive, realistic and up-to-date plan for the remaining work which he is obliged to do by the ECC and for which serious sanctions apply in the event that he fails to do so. The side-benefit of these requirements is to eliminate disputes over the applicability of tendered rates for pricing variations and to reduce the likelihood of disputes over the time effect of those same variations.

4.3.3 Clear division of function and responsibility

The ECC recognises that if risk is placed on a Party to the Contract he is motivated to minimise its effect and use risk allocation to encourage good management in the parties most likely to be able to respond.

For example, the *Contractor* is traditionally assumed to have inspected the Site and carried out his own site investigation. This does not motivate the *Employer* to do sufficient site investigation to establish the effect of ground conditions on construction cost. The ECC does not state that the *Employer* should do more site investigations, as this would have little impact. Instead it is stated that the *Contractor* is to assume that the ground conditions will be as they are described to him in the tender documents (the Site Information).

Consequently, if only minimal site investigation has been done, the *Contractor* could base his price on an erroneous view of the sub-surface conditions. As this will increase the *Employer*'s risk of later programme delays and extra cost, the *Employer* is more strongly motivated to investigate sufficiently. The side-benefit of course is that if the *Employer* is motivated to do a comprehensive site investigation, the likelihood of disputes arising over changed conditions and their effect on the *works* must be reduced.

Staying with site investigations and physical conditions, it was stated earlier that disputes rooted in different interpretations placed on the Site Information are rarely black-and-white affairs. This in part is due to the difficulty in defining precisely the boundary between the physical condition risk carried by the *Contractor* and that by the *Employer*. This difficulty is exacerbated by the use in construction contracts of such terms as 'those physical conditions which could not reasonably have been foreseen by an experienced *Contractor*' to describe the risk carried by the *Employer*. A simple example of the problem will illustrate the difficulty.

> The Site Information includes three borehole logs, which indicate that in three separate locations the depth of the existing topsoil at the Site was 100 mm, 200 mm and 350 mm respectively. Is the *Contractor* deemed to have included in the price for excavating existing topsoil across the Site up to 350 mm thick, an average of 217 mm thick across the Site or some other permutation of the numbers depending on the precise location of the boreholes?
>
> The ECC2 Guidance Notes have recognised the potential for such issues to give rise to disputes and so they suggest the inclusion in the contract of 'boundary limits' between the risks carried by the *Employer* and the *Contractor*; that is, to state what tenderers should allow for in their tenders.
>
> In our topsoil example this could be done by stating in the Works Information that the *Contractor* shall be deemed to have allowed in the tendered total of the Prices for excavating a prescribed volume of topsoil with a thickness in the range 100–350 mm. Additionally, a small tolerance say ±5% could be stated to apply to the prescribed volume to avoid compensation events for insignificant changes in quantity. Tenderers will then be able to tender on a common basis knowing that they must allow in their pricing for the occurrence of physical conditions within the stated boundary limits. The same principles can of course be applied to such physical conditions as soil characteristics, level of rock–soil interface, groundwater levels, permeability limits and overbreak in rock excavation.

4.3.4 Reducing disputes in the Works Information and Site Information

This section examines how the ECC attempts to head off potential disputes centring on deficient/inadequate Works and Site Information, looking individually at cases where:

- the Works Information is unclear as to what is to be done,
- there exists ambiguity or inconsistency in or between the documents comprising the Works Information,
- the *Contractor* contends that the Works Information requires him to do something which is illegal or impossible,
- there exist inconsistencies within the Site Information (including the information referred to in it).

4.3.4.1 Works Information descriptions

This situation is where the Works Information does not provide a full description of what is to be done by the *Contractor* or does not describe adequately the constraints under which the work is to be done.

For example, on a contract where the *Employer* is responsible for all design, the Works Information might be silent on the subject of what tests are necessary to verify the quality of a particular component or the construction tolerances applicable. In such instances it falls to the *Project Manager* to remedy this deficiency as part of his duty to ensure the completed works meet the *Employer*'s objectives in terms of quality. Clause 14.3 of the ECC gives the *Project Manager* the power to instruct a change to the Works Information[178] and in the example referred to, the *Project Manager* would instruct the *Contractor* as to what tests were necessary, or the construction tolerance that was applicable. Such an instruction, being a change to the Works Information, is of course a compensation event[179] and the *Contractor* would therefore have a time and monetary entitlement, which if nothing else, emphasises the importance of getting the Works Information right in the first instance.

> Get the Works Information right!
>
> Deficiencies in the Works Information may lead to compensation events.

4.3.4.2 Conflicts within the Works Information

The second situation is where separate parts of the Works Information might in themselves be clear as to what is to be done but unfortunately, conflict with one another, for example the same reinforcement bars called up as two different diameters in the reinforcement schedules and on the reinforced concrete detail drawings. The ECC does not utilise a hierarchical arrangement of the contract documents, giving precedent to those higher up the hierarchy to resolve such ambiguities. Instead, where such ambiguities and inconsistencies exist in or between the documents comprising the Works Information, the ECC places the responsibility on the *Project Manager* to give an instruction resolving the ambiguity or inconsistency, for example by advising the *Contractor* what the *Employer*/designer actually requires.[180] If the resolution of the ambiguity or inconsistency requires an instruction, which changes the Works Information, the 'contra proferentum' rule applies which interprets any ambiguous or inconsistent statements in a contract against the party responsible for their preparation.[181]

[178] In ECC3 the words 'or a Key Date' have been added.
[179] Clause 60.1(1).
[180] Clause 17.1.
[181] ECC2 clause 63.7; ECC3 clause 63.8 – also includes 'Key Dates'.

> To put this in a practical context, using the reinforcement example above, if the reinforcement schedule showed 32 mm bars and the reinforcement detail drawing for the same bars indicated a diameter of 40 mm, then if the *Project Manager* confirms the latter as being required, the effect of the ensuing compensation event would be assessed as if the Prices and the Completion Date[182] were for the interpretation most favourable to the *Contractor*; that is, he would be deemed to have allowed in his original tender price for the smaller bars.

> **The ECC has no hierarchy of documents.**
>
> The contra proferentem rule applies against the *Employer* if the Works Information contains conflicting information, since ambiguities and inconsistencies in wording are construed against the drafter.

4.3.4.3 Works Information requires an illegal action

On to the third situation where the *Contractor* notifies the *Project Manager* that he considers that the Works Information requires him to do something which is illegal (e.g. flout the Building Regulations) or impossible[183] (e.g. to construct a bored tunnel causing absolutely no settlement). If, having reviewed the situation, the *Project Manager* agrees, he gives an instruction changing the Works Information appropriately. This of course would be a compensation event.[184]

4.3.4.4 Inconsistencies within the Site Information

Finally, where the Site Information contains inconsistencies within itself or between it and other information referred to in it, for example one part of the Site Information shows the Site to be clear of all buried services while another part shows a high-pressure gas main crossing the Site. Again the 'contra-proferentem' rule applies, clause 60.3[185] stating that 'the *Contractor* is assumed to have taken into account the physical conditions more favourable to doing the work'; that is, in this case to have assumed the Site to be clear of all buried services.

> **The contra proferentem rule applies to the Site Information.**

4.3.5 Conclusion

All of the above are examples of 'self-help' remedies contained within the ECC, facilitating the removal of the uncertainty as to what is to be done in particular circumstances and in doing so reducing the potential for disputes.

Before concluding this section, it is worth countering the concerns from many quarters that by using the ECC there is a risk that claims on contractual matters will increase because the contract is relatively untried, the language is unfamiliar and the contract has never been tested in the courts. The ECC had more legal checking before publication than any of the traditional standard forms preceding it. It was drafted to eliminate known problems associated with traditional contracts, which have come before the courts. All the well-known court cases which hinged upon the wording of a traditional contract have been taken into account in drafting the ECC so that the same matter could not come up again when the ECC was used. In addition, the fact that it has not been tested in court is extremely positive – it means that the contract has not been so troublesome that a court has had to resolve conflicts.

We will now consider in the final section of this chapter the situation where, despite all the efforts of both parties, and the self-help remedies contained within the ECC, a dispute actually arises.

---

[182] ECC3 clause 63.8 extends this clause to cover 'Key Dates'.
[183] ECC2 clause 19.1; ECC3 clause 18.1 – changes emphasis from 'becomes aware' to 'considers'.
[184] Clause 60.1(1).
[185] ECC3 clauses 60.2 and 60.3 have been merged into clause 60.2 and the wording has been amended to include 'ambiguities' as well as inconsistencies

4.4 Dispute resolution under the ECC

4.4.1 General The coming into effect of the Housing Grants, Construction and Regeneration Act 1996 for contracts let after 1 May 1998 gives each party to a contract the right to refer disputes to adjudication. Since ECC2 caters for adjudication, those contracts that do not fall within the definition of a construction contract as stated in the Act could still use the ECC adjudication procedure to resolve disputes. The published NEC ECC secondary Option Y(UK)2 should be chosen and used for contracts under ECC2 that do fall within the definition of a construction contract as stated in the Act. ECC3 caters for adjudication in dispute resolution Options W1 (used except when the HGCR Act 1996 applies) and W2 (used when the HGCR Act 1996 applies). Option Y(UK)2 in ECC3 refers to payment rather than adjudication.

> The ECC caters for adjudication on contracts that **do not** fall within the definition of a construction contract or are outside the UK.
>
> Secondary Options Y(UK)2 should be chosen for contracts that fall within the definition of a construction contract.

4.4.2 Adjudication: pre-HGCR Act 1996 Until the enactment of the Housing Grants, Construction and Regeneration Act 1996, the only means of dispute resolution in many forms of contract was arbitration or litigation which in recent years have become both time-consuming and expensive. While the ECC recognises the need to have an ultimate means for dispute resolution it introduces an intermediate stage of independent dispute resolution, in the form of adjudication. It is the intention that all disputes arising under or in connection with the ECC that could not be settled by the *Project Manager* and the *Contractor* should be dealt with and settled by the *Adjudicator* who is appointed jointly by the *Employer* and the *Contractor*. The person appointed as the *Adjudicator* is named in the Contract Data part one and preferably engaged under the NEC Adjudicator's Contract under which the Parties indemnify the *Adjudicator* against claims, etc. and agree to share his fees equally, regardless of his decision.

The ECC identifies three classes of dispute, which can be submitted to and settled by the *Adjudicator*, namely[186]

- disputes about an action of the *Project Manager* or *Supervisor* (which includes disputes about the outcome of such action, e.g. the decision resulting from the action),
- disputes about the lack of action by the *Project Manager* or the *Supervisor*,
- disputes about any other matters (arising under or in connection with the contract).

The first two of the above classes of dispute are likely to be most common in practice and the adjudication procedure laid down in the ECC for such disputes follows these steps:

(1) The disputed action, lack of action or other matter occurs.
(2) Within four weeks of the disputed action/inaction occurring, the *Contractor* notifies the dispute to the *Project Manager*.
(3) Between two and four weeks after the *Contractor*'s above notification of the disputed matter, and assuming the *Project Manager* has not taken or amended the action, the *Contractor* submits the dispute to the *Adjudicator* including with his submission any relevant information he wishes to be considered by the *Adjudicator*. A copy of the submission must also be provided to the *Employer*.

The *Employer* (in practice more probably the *Project Manager*) should within four weeks of the *Contractor*'s submission of the dispute, submit to the

---

[186] ECC3 dispute resolution Option clause W1 (used except when the HGCR Act 1996 applies) adds a fourth reason: 'A quotation for a compensation event which is treated as having been accepted'.

Adjudicator (copy to the *Contractor*) any information upon which he relies by way of a response. The *Contractor* during the same four-week period may also submit to the *Adjudicator* any further information not included with his original submission. This four-week period is vital to ensuring that the *Adjudicator* has all the relevant information to enable him to put himself in the position of the *Project Manager* when the disputed action was taken or not taken as the case may be. It is therefore intended that this period be used by the Parties to assist the *Adjudicator* in reaching as rapid a decision as is reasonably possible. The *Adjudicator* during this time could call for further information he considers necessary to assist him in reaching a decision. The four-week period may be extended by the *Adjudicator* with the agreement of the Parties.

Within four weeks of the end of the period for providing information, the *Adjudicator* notifies his decision together with his reasons. Again the four-week period may be extended by the *Adjudicator* with the agreement of the Parties.

Some examples of decisions that may result from the Adjudication process are now considered in order to illustrate how such decisions could be implemented.

> If the *Adjudicator* decides in the *Contractor*'s favour but it is too late for the action or inaction to be implemented, he will deal with the matter by deciding the effect on the Prices and Completion Date using the same assessment procedure that is used for compensation events.
>
> In other circumstances, it might be appropriate for the *Adjudicator* to change the disputed action or inaction. For instance, where the *Contractor* disputes the existence of a Defect, which has been notified by the *Supervisor*, the *Adjudicator* might decide in the *Contractor*'s favour. If so, the *Contractor* would be relieved of any obligation to correct the Defect, if corrective work had not started. If the alleged defect had already been 'corrected', the *Adjudicator* would decide on the financial and time effects. However, if despite the *Adjudicator*'s decision, the *Project Manager* still requires additional or remedial work, then he may instruct such work as a change to the Works Information. This would be a compensation event.
>
> If the dispute concerned an amount due as certified by the *Project Manager* and the *Adjudicator* decides that the amount certified was incorrect, the *Project Manager* will be required to make a correction in the next certificate and include interest as required by clause 51.3 of the ECC.
>
> If the *Adjudicator* disagrees with the *Project Manager*'s assessment of delay to the Completion Date, he will overrule the *Project Manager*'s decision and the Completion Date will be set in accordance with what the *Adjudicator* decides. If, however, it is too late to allow the *Contractor* to revise his programme, the *Adjudicator*'s decision will be in respect of the effect on the Prices only.

A dispute is settled when the *Adjudicator* notifies the Parties and the *Project Manager* of his decision. Unless and until there is such a settlement, the Parties and the *Project Manager* proceed as if the matter were not disputed. The *Adjudicator*'s decision is final and binding unless and until revised by the *tribunal*.

The *Adjudicator* can, by agreement of the Parties, join a dispute between *Employer* and *Contractor* in the main contract, which also involves a dispute between the *Contractor* and his Subcontractors.

> The *Adjudicator*'s decision is final and binding unless revised by *tribunal*.
>
> A dispute cannot be referred to *tribunal* unless it has first been referred to adjudication.

4.4.3 Adjudication: post-HGCR Act 1996

Part II of the Housing Grants, Construction and Regeneration Act took effect from 1 May 1998 and all construction contracts entered into after that date are subject to it.

So, how is it relevant to the ECC2, which after all, already provides for adjudication as a first-stage dispute resolution process? Well, to avoid the imposition of the 'fallback' provisions of the Scheme for Construction Contracts, all construction contracts as defined under the Act must comply with the Act. While the ECC does provide for adjudication, it does not meet all of the essential points of compliance contained in the Act.

Option Y(UK)2[187] takes into account the Housing Grants, Construction and Regeneration Act 1996 and rectifies the original deficiency. The reason for this matter being dealt with by a new secondary Option (as opposed to amendments to the core clauses) is explained by the fact that the NEC intends its contract to be used internationally and in such circumstances it will be outside the embrace of the Act. The relevant clauses of the new secondary Option for adjudication (some clauses address payment issues) are clauses Y2.5, Y2.6, Y2.7 and Y2.8.

The problem the ECC2 faced over the adjudication provisions of the Act was the right that is given to each party to refer disputes arising under the contract to adjudication at any time. The NEC originally sought to time-bar the notification of disputes as part of its philosophy of flushing out and resolving issues at the earliest opportunity, thereby underpinning its objectives of certainty of outcome for the parties. It sought to do this by preventing disputes being referred to the *Adjudicator* outside a four-week period calculated from when the *Contractor* became aware of the dispute. To prevent the Act undermining its principles and objectives, secondary Option Y(UK)2 seeks to introduce a preliminary stage to the notification of a dispute by introducing the concept of '**dissatisfaction**'.

Clause Y2.5 provides that if the *Contractor* is dissatisfied with an action/inaction of the *Project Manager*, he notifies his dissatisfaction to the *Project Manager* no later than four weeks after he became aware of the action/inaction. Within two weeks of such notification, the *Contractor* and the *Project Manager* attend a meeting to discuss and seek to resolve the matter. The 'clincher' is that the Parties agree that no matter shall be a dispute unless a notice of dissatisfaction has been given and the matter has not been resolved within four weeks. In summary therefore, the ECC2 introduces a procedural hurdle denying the Parties the freedom to refer a dispute to adjudication at any time by redefining a dispute as an unresolved matter which has been the subject of a notice of dissatisfaction and by forcing the participants to follow a procedure that stays true to the original objectives and principles of the ECC2.

It is unlikely that this amendment is acceptable, however, and it is recommended to users of Option Y(UK)2 in ECC2 that they seek legal advice if they wish to consider deleting clause 90.4 of Y2.5 from contracts.

The remaining amendments introduced by secondary Option Y(UK)2 concern themselves with ensuring the time-scales for the adjudication process comply with the Act and addressing the remaining essential points of compliance described above. They do not, to any significant extent, affect the adjudication process described in the ECC2.

4.4.4 Adjudication in ECC3

The drafters of the NEC have taken the opportunity to rationalise and incorporate the requirements of the Housing Grants, Construction and Regeneration Act 1996 into the body of the contract for the third edition of the ECC.

In ECC2 there was often debate that the provisions of Y(UK)2, which outlined the amendments to be made to the core clause to comply with the Act, did in fact not achieve its objective.

---

[187]ECC2 only since Y(UK)2 in ECC3 does not deal with adjudication.

The provision in ECC2 for adjudication in core clause section 9, clauses 90, 91, 92 and 93 have been removed and replaced in ECC3 with Dispute Resolution Procedure Options W1 and W2. The thinking behind this change is very simple. The NEC is an international contract and as such a national requirement for the HGCR Act 1996, which is a UK-specific requirement, should not be contained within the core clauses of the contract. Therefore Option W1 is to be used except in the UK when the Housing Grants, Construction and Regeneration Act 1996 applies.[188]

Option W1 applies to all non-UK contracts and those contracts in the UK which fall outside of the definition of a construction contract for the purposes of the HGCR Act 1996.

Option W2 is to be used in the UK when the Housing Grants, Construction and Regeneration Act 1996 applies and must be used with secondary Option Y(UK)2. Y(UK)2 ensures compliance with the requirements of sections 110 (Dates for Payment), section 111 (Notice of intention to withhold payment), section 112 (Right to suspend performance for non payment) and section 116 (Reckoning of time periods).

The ECC was the first contract to make provision for adjudication because the drafters of the NEC wanted disputes to be resolved quickly. They also wanted the raising of disputes to be time-barred and Option W1 maintains this principle in clause W1.3 (2).[189]

The very first entry in Contract Data part one requires the user to insert which dispute resolution procedure Option applies to the contract: Option W1 or Option W2.

Table 4.1 shows how ECC3 complies with the requirement of the HGCR Act 1996.

4.4.5 The *tribunal* The *tribunal* is the second formal level of dispute resolution in the ECC. The *tribunal* is chosen by the *Employer*[190] and would normally be either arbitration or the courts.

For contracts which do not fall under the definition of a construction contract under the HGCR Act 1996 in ECC2 or if Option W1 is chosen in ECC3 then a dispute cannot be referred to the *tribunal* unless it has first been referred to the *Adjudicator*[191] and tribunal proceedings cannot be commenced until Completion of the whole of the *works* has taken place (or termination).[192]

In ECC3 the optional statements given in Contract Data part one have been extended not only to state what the arbitration procedure is to be but also:

- The place where arbitration is to be held is . . . .
- The person or organisation who will choose an arbitrator
  - if the Parties cannot agree a choice, or
  - if the arbitration procedure does not state who selects an arbitrator, is . . . .

(Note: refer to Option W1 clause W1.4(5) and Option W2 clause W2.4(4))

---

[188]Note not all construction projects in the UK fall under the HGCR Act 1996 – see sections 104 to 107 of the Act.
[189]ECC2 clause 91.1.
[190]By an appropriate insertion in part one of the Contract Data.
[191]ECC2 clause 93.1; ECC3 clause W1.4 of Option W1.
[192]This is not a requirement in ECC3.

Table 4.1 ECC3 Option W2 dispute resolution procedure – compliance with HGCR Act 1996

Housing Grants, Construction and Regeneration Act 1996	ECC3 Option W2 Clause reference
Section 108 Adjudication – Right to refer disputes to adjudication	
108(1) A party to a construction contract has the right to refer a dispute under the contract for adjudication under a procedure complying with this section. For the purpose 'dispute' includes any difference.	
108(2) The contract shall	
(a) enable a party to give notice **at any time** of his intention to refer a dispute to adjudication	W2.1 'A party may refer a dispute to the *Adjudicator* **at any time**.'
(b) provide a timetable with the object of securing the appointment of the adjudicator and referral of the dispute to him **within 7 days** of such notice	W2.3(1) 'Within **three days** of the receipt of the notice of adjudication the adjudicator notifies the parties • that he is able to decide the dispute in accordance with the contract, or • that he is unable to decide the dispute, in which case he resigns. W2.3(2) 'Within **seven days** of giving notice of adjudication the Party referring the dispute to the *Adjudicator* • refers the dispute to the *Adjudicator*, • provides the *Adjudicator* with the information on which he relies, including any supporting documents and • provides a copy of the information he has provided to the *Adjudicator* to the other Party. W2.2(3) 'If the *Adjudicator* is not identified in the Contract Data or if the *Adjudicator* resigns or becomes unable to act • the Parties choose an adjudicator jointly, or • a Party may ask the *Adjudicator* nominating body to choose an adjudicator **within 4 days** of request.
(c) require the adjudicator to reach a decision within 28 days of referral or such longer period as is agreed by the parties after the dispute has been referred	W2.3(8) 'The *Adjudicator* decides the dispute and notifies the Parties and the *Project Manager* of his decision and his reasons within twenty eight days of the dispute being referred to him. This twenty eight day period may be extended by up to fourteen days with the consent of the referring Party, or by any period agreed by the Parties.'
(d) allow the adjudicator to extend the period of 28 days by up to 14 days, with the consent of the party by whom the dispute was referred	W2.3(8) (see above).
(e) impose a duty on the adjudicator to **act impartially** and	W2.2(2) 'The *Adjudicator* **acts impartially** and decides the dispute as an independent adjudicator and not as an arbitrator.'
(f) enable the adjudicator to take the **initiative in ascertaining the facts and the law.**	W2.3(4) 'The Adjudicator may • review and revise any action or inaction of the *Project Manager* or *Supervisor* related to the dispute, • take **the initiative in ascertaining the facts and the law relevant to the dispute**, • require a Party to provide further information related to the dispute and • issue any other instruction to the Parties he considers necessary in order to reach his decision.

Table 4.1 *Continued*

Housing Grants, Construction and Regeneration Act 1996	ECC3 Option W2 Clause reference
108(3) The contract shall provide that the decision of the adjudicator is **binding** until the dispute is finally determined by legal proceedings, by arbitration (if the contract provides for arbitration or the parties otherwise agree to arbitration) or by agreement. The parties may agree to accept the decision of the adjudicator as **finally determining the dispute**.	W2.3(11) 'The *Adjudicator*'s decision is **binding** on the Parties unless and until revised by the *tribunal* and is enforceable as a matter of contractual obligation between the Parties and not as an arbitral award. The *Adjudicator*'s decision is **final and binding** if neither Party has notified the other within the times required by this contract that he is dissatisfied with a matter decided by the *Adjudicator* and intends to refer the matter to the *tribunal*.'
108(4) The contract shall also provide that the adjudicator is not liable for anything done or omitted in the discharge or purported discharge of his functions as adjudicator unless the act or omission is in **bad faith**, and that any employee or agent of the adjudicator is similarly protected from liability.	W.2.2(5) 'The *Adjudicator*, his employees and agents are not liable to the Parties for any action or failure to take action in an adjudication unless the action or failure to take action was in **bad faith**.'
108(5) If the contract does not comply with the requirements of subsections (10) to (4), the adjudication provisions of the Scheme for Construction Contracts apply.	Failure to comply will invoke the provisions of the Scheme for Construction Contracts. This is not an actual clause in Option W2, but the law in ECC3.
Housing Grants, Construction and Regeneration Act 1996[a]	Option Y(UK)2: the Housing Grants, Construction and Regeneration Act 1996
Section 116 Reckoning periods of time (3) Where the period would include Christmas Day, Good Friday or a day which under the Banking and Financial Dealings Act 1971 is a bank holiday in England and Wales or, as the case may be, in Scotland, that day is excluded.	Y2.1(2) Definitions 'A period of time stated in days is a period calculated in accordance **with Section 116 of the Act**.'

Section 110 Dates for payment
(1) Every construction contract shall
 	
   (a) provide an adequate mechanism for determining **what payments become due under the contract, and when**, and
   (b) provide for a **final date for payment** in relation to any sum which becomes due.
 	
The parties are free to agree how long the period is to be between the date on which a sum becomes due and the final date for payment.
	Y2.2 Dates for Payment
	'**The date on which, a payment becomes due is seven days after the assessment date**' (the latest date for payment under clause 51.1).
	
	'The **final date for payment is fourteen days or a different period for payment if stated in Contract Data** after the date on which payment becomes due' (the latest date for payment under clause 51.2).

(2) Every construction contract shall provide for the giving of notice by a party not later than five days after the date on which payment becomes due from him under the contract, or would have become due if
 	
   (a) the other party had carried out his obligations under the contract, and
   (b) no set-off or abatement was permitted by reference to any sum claimed to be due under one or more other contracts
 	
specifying the amount (if any) of payment made or proposed to be made, and the basis on which that amount was calculated.
	Clause 51.1 of ECC3.

(3) If or to the extent that a contract does not contain such provision as is mentioned in subsection (1) or (2), the relevant provisions of the Scheme for Construction Contracts apply.	

Table **4.1** *Continued*

Housing Grants, Construction and Regeneration Act 1996[a]	Option Y(UK)2: the Housing Grants, Construction and Regeneration Act 1996

Section 111 Notice of intention to withhold payment
(1) A party to a construction contract **may not withhold payment after the final date for payment of a sum due under the contract unless he has given an effective notice of intention to withhold payment.**
 	
The notice mentioned in section 110(2) may suffice as a notice of intention to withhold payment if it complies with the requirements of this section.
 	
(2) To be effective such notice **must specify**
 	
  (a) **the amount proposed to be withheld and the ground for withholding payment,** or
  (b) **if there is more than one ground, each ground and the amount attributable to it,**
 	
and must be given not later than the prescribed period before the final date for payment.
 	
(3) The parties are free to agree what the **prescribed period** is to be.
 	
In the absence of such agreement, the period shall be that provided by the Scheme for Construction Contracts.
	Y2.3 Notice of intention to withhold payment
	'If either Party intends to withhold payment of an amount due under this contract, he notifies the other Party not later than seven days **(the prescribed period)** before the final date for payment **by stating the amount proposed to be withheld and the reason for withholding payment. If there is more than one reason, the amount for each reason is stated.**
	
	A Party does not withhold payment of an amount due under this contract unless he has notified his intention to withhold payment as required by this contract.

Section 112 Right to suspend performance for non-payment (1) Where a sum due under a construction contract is not paid in full by the final date for payment and no effective notice to withhold payment has been given, the person to whom the sum is due **has the right (without prejudice to any other right or remedy) to suspend performance of his obligations under the contract** to the party by whom payment ought to have been made ("the party in default"). (2) The right may not be exercised without first giving to the party in default at least seven days' notice of intention to suspend performance, stating the ground or grounds on which it is intended to suspend performance. (3) The right to suspend performance ceases when the party in default makes payment in full of the amount due. (4) Any period during which performance is suspended in pursuance of the right conferred by this section shall be disregarded in computing for the purposes of any contractual time limit the time taken, by the party exercising the right or by a third party, to complete any work directly or indirectly affected by the exercise of the right. Where the contractual time limit is set by reference to a date rather than a period, the date shall be adjusted accordingly.	**Y2.3 Suspension of performance** 'If the *Contractor* exercises his right under the Act to suspend performance it is a compensation event.'

[a]Extract from Housing Grants, Construction and Regeneration Act 1996.

4.5 Adjudication – general comments and observations

The drafters of the ECC included the adjudication process as an independent third-party review of disputes. Many commentators call it a 'quick and dirty' process.

It is interesting to note how the adjudication process is seen. There are two schools of thought:

(1)   the adjudication is in the Contract to be used or
(2)   adjudication is a failure (because the dispute has not been resolved prior to the adjudication process).

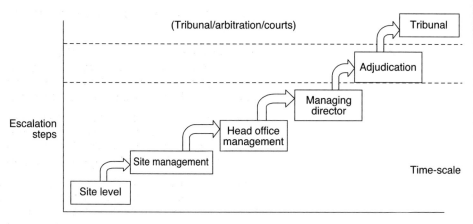

Fig. 4.1. Dispute escalation

In relation to item (2), many 'enlightened' clients see adjudication as a failure and introduce through Option Z (additional conditions) clauses and procedures that deal with dispute resolution on a contract.

This usually revolves around the idea of dispute being resolved at the lowest level possible and there being an 'escalation' process for the more difficult issues. Only after the parties have exhausted this in a structured time-scale, is adjudication used.

This type of arrangement is often found in partnering and frameworking arrangements.

Figure 4.1 shows an example of the 'steps' in the escalation of a dispute.

4.5.1 'Star Chambers' and the like

A 'Star Chamber' is simply an interim dispute resolution process according to which, if the parties fail to agree at the lowest level, then the dispute will ultimately be resolved by the 'Star Chamber'. The chamber comprises the managing directors or other senior executives of the two parties. Some commentators consider that such a chamber also needs an independent representation wherever the 'Star Chamber' is unable to agree a decision.

If the issue gets as far as the 'Star Chamber', each party will be given an opportunity to resolve the issue at the lowest level, usually to a prescribed time-scale.

After having been given time to resolve the issue at the lowest level, each party will be given a set period, say two weeks, in which to compile the facts as they see them. Each side then also presents this information in a 15/20-minute presentation. The 'Star Chamber' members then deliberate on the issue and give their decision.

The process relieves the lower levels of management from making decisions on disputed items, and gets them resolved so that they can concentrate on doing their role on the project (adding value), rather than being distracted from their role.

The authors' own experience is that having to present your argument/reasoning on why something is or is not a compensation event is daunting, especially if your audience is your company's managing director or client.

It is interesting to note that such a procedure does encourage resolution at the lowest level. People tend not to want to appear unable to resolve issues perceived as being confrontational.

4.5.2 Good information and records – how the *Adjudicator* will judge the information

It is essential to keep good records and information in a structured way as required by the ECC.

The *Adjudicator* will review the actions/inaction of the Parties based upon the information and records existing at the time the issue/dispute arose.

The added benefit of good information and records is that if/when a dispute arises you will not be involved in hours and days of documentation retrieval. This

is especially true where IT has been harnessed to control the administration of change.

4.5.3 The *Adjudicator*

The *Employer* will insert in Section 1 General in Contract Data part one, the name of the *Adjudicator*. Before naming the *Adjudicator*, the *Employer* should check that the *Adjudicator* has no conflicts of interest with the Parties and that he has the relevant experience, qualifications and competency.

Some *Employers* have a list of *Adjudicators* from which to choose,[193] others simply leave it open for the *Adjudicator* to be selected by 'the President for the time being of the Institution of Civil Engineers' or other professional bodies. Most of the professional bodies keep a register of approved adjudicators.

It is essential to name the *Adjudicator* in the contract prior to contract execution. If Parties are already in dispute, it may become unlikely that they will agree the name of the *Adjudicator* for contracts where 'to be agreed' has been inserted against the *Adjudicator* in the Contract Data.

4.5.3.1 When should the Adjudicator become involved?

The *Adjudicator* only becomes involved when a dispute arises. He will then be appointed jointly by the *Employer* and *Contractor* under the NEC Adjudicator's Contract.

His fees are shared between the Parties regardless of his decision and regardless of which Party refers the dispute.

4.5.3.2 Who should be the Adjudicator?

Construction projects are complex and disputes can arise on technical or commercial issues. On the technical side, the dispute may involve specialist work such as geotechnical, or specialist engineering systems which need to meet prescriptive performance tests.

Therefore even the most experienced professional acting as an *Adjudicator* is unlikely to be an expert or knowledgeable on everything.

Some people suggest having a number of adjudicators named to cover engineering and commercial aspects. However, this should be unnecessary since the named *Adjudicator* should have the relevant experience/competence to draw upon the technical assistance of others.

> The *Adjudicator*'s name should be inserted in Contract Data part one.
>
> The *Adjudicator* only becomes involved when a dispute arises.
>
> Appointed jointly by the *Employer* and the *Contractor* on the NEC Adjudicator's Contract.
>
> Adjudication fees are shared equally irrespective of his decision.

> The ECC was the first contract to have adjudication built in. However, in the UK the contract was somewhat hijacked by legislation in the form of the Housing Grants, Construction and Regeneration Act 1996 (HGCR Act 1996), since which time there has been much debate about whether or not the ECC is compliant with the HGCR Act 1996.
>
> A further consideration for the NEC drafting panel is the fact that the NEC is an international contract and the requirements of the HGCR Act 1996 is an issue for UK contracts only.
>
> To overcome these problems, new secondary Options W1 and W2 have been introduced, which relate to the dispute resolution procedure to be adopted.
>
> Secondary Option W1 retains the intent for disputes to be notified within four weeks after which they become time-barred.
>
> *Cont'd*

---

[193] Note that having an *Adjudicator* on a retainer basis defeats the purpose of an independent arbiter of disputes.

> Secondary Option W2 should be used in conjunction with secondary Option clause Y(UK)2 to make a UK-based contract compliant with the requirements of the HGCR Act 1996.
>
> Drafters of international contracts need to remember that Option Y(UK)2 is a reflection that the NEC is an international, not purely a national contract and that other countries may have their own legislation and requirements.

4.6 NEC 3rd Edition

The following table indicates the changes made in ECC3 in relation to dispute resolution.

ECC2 clause	ECC3	Comments
Y(UK)2	Y(UK)2 Totally rewritten outlining the essentials requirements of the HGCR Act 1996 sections. Y(UK)2 ensures compliance with the requirements of section 110 (Dates for Payment), section 111 (Notice of intention to withhold payment), section 112 (Right to suspend performance for non payment) and section 116 (Reckoning of time periods). Used for UK construction contracts where the contract falls under the definition of a construction contract under sections 104 to 107 of the Act. It also requires new secondary Option W2 to be chosen which replaces clauses 90, 91, 92 and 93 which dealt with adjudication in ECC2.	
90, 91, 92 and 93	Options W1 and W2 Dispute Resolution Procedure Reference to Adjudication has been removed totally from core clause section 9. In particular clauses 90, 91, 92 and 93 of NEC2 have been deleted and replaced with new Options W1 and W2 Dispute Resolution Procedure. Option W1 retains the drafters' intent of matters for adjudication being time-barred. Not to be used for UK contracts.	The redrafted secondary Options W1 and W2 are essentially the same as clauses 90, 91, 92 and 93 in ECC2. There is also some tidying up of the wording and the inclusion of some new requirements to make ECC3 fully Act compliant.

Figure 4.2 describes how the ECC user ensures HGCR compliance in ECC3.

The following table outlines the most significant changes made in Option W1 which is a redraft of ECC2 clauses 90, 91, 92 and 93.

Dispute resolution procedure Option W1	
Dispute resolution W1.1	The same as clause 90.1 in ECC2; the word 'referred' is replaced by 'submitted'.
The *Adjudicator* W1.2(1)	New requirement. 'The Parties appoint the *Adjudicator* under the NEC Adjudicator's Contract current at the *starting date*.' This clause reinforces the requirement to appoint the *Adjudicator* prior to the commencement date of the contract. This is a common problem on many contracts where the insertion of the name of the *Adjudicator* is left blank.
W1.2(5)	New clause: 'The *Adjudicator*, his employees and agents are not liable to the Parties for any action or failure to take action in an adjudication unless the action or failure to take action was in bad faith.' This new clause reflects the common law situation relating to adjudicators.
W1.3 Adjudication table	New dispute item added. 'A quotation for a compensation event which is treated as having been accepted.' May be referred by the *Employer*. This item has been added to reflect the changes made in ECC3 new clause 62.6 whereby the *Project Manager*'s failure to reply to a quotation is deemed to be acceptance.

Dispute resolution procedure Option W1 *Continued*

W1.3(2)	'The times for notifying and referring a dispute may be extended by the *Project Manager* if the *Contractor* and the *Project Manager* agree to the extension before the notice or referral is due. The *Project Manager* notifies the extension that has been agreed to the *Contractor*. If a disputed matter is not notified and referred within the times set out in this contract, neither Party may subsequently refer it to the *Adjudicator* or the *tribunal*.'
	By agreement the Parties may extend the times for notifying and referring a dispute before the notice or referral is due.
	The second part of this clause provides a time bar on disputes being referred to the *Adjudicator* or *tribunal*.
W1.3(11)	New requirements.
	'The *Adjudicator* may, within two weeks of giving his decision to the Parties, correct any clerical mistake or ambiguity.'
Review by Tribunal W1.4(1)	New requirement.
	'A Party does not refer any dispute under or in connection with this contract to the tribunal unless it has first been referred to the *Adjudicator* in accordance with this contract'. This clause makes it clear that all disputes must firstly go to the *Adjudicator* as the first point of dispute resolution.
W 1.4(5)	New requirement.
	'If the *tribunal* is arbitration, the *arbitration procedure*, place where the arbitration is to be held and the method of choosing the arbitrator are those stated in the Contract Data.' The information required by this clause will need to be inserted into Contract Data part one.
W1.4(6)	New requirement.
	'A party does not call the *Adjudicator* as a witness in the *tribunal* proceedings'.

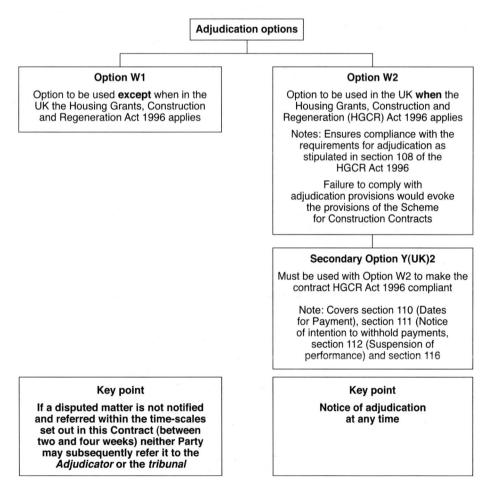

Fig. 4.2. NEC3 adjudication options

The following table outlines the most significant changes made in Option W2

Dispute resolution procedure Option W2	
Dispute resolution W2.1(1)	The same as clause 90.1 in ECC2; the word 'referred' is replaced by 'submitted'. Also includes the following additional sentence:
	'A Party may refer a dispute to the *Adjudicator* at any time.'
	This clause reflects the fact that in order to comply with the HGCR Act 1996 a dispute must be able to be referred to the *Adjudicator* at any time. Section 108(2) of the Act.
W2.1(2)	'In this option, time periods stated in days exclude Christmas Day, Good Friday and bank holidays.'
	This provides compliance with section 116 of the HGCR Act 1996. Formerly included in ECC2 in secondary Option Y(UK)2 clause Y2.1.
The *Adjudicator* W2.2(1)	New requirement.
	'The Parties appoint the *Adjudicator* under the NEC Adjudicator's Contract current at the *starting date*.'
	This clause reinforces the requirement to appoint the *Adjudicator* prior to the commencement date of the contract. This is a common problem on many contracts where the insertion of the name of the *Adjudicator* is left blank.
W2.2(5)	New clause: 'The *Adjudicator*, his employees and agents are not liable to the Parties for any action or failure to take action in an adjudication unless the action or failure to take action was in bad faith.'
	This new clause reflects the common law situation relating to adjudicators.
The *adjudication* W2.3(1)	'Before a Party refers a dispute to the *Adjudicator* he gives a notice of adjudication to the other Party with a brief description of the dispute and the decision that he wishes the *Adjudicator* to make. If the *Adjudicator* is named in the Contract Data the Party sends a copy of the notice of adjudication to the *Adjudicator* when it is issued. Within three days of the receipt of the notice of adjudication the *Adjudicator* notifies the parties
	• that he is able to decide the dispute in accordance with the contract, or • that he is unable to decide the dispute, in which case he resigns.
	If the *Adjudicator* does not notify either of these alternatives within 3 days of the issue of the notice of adjudication either Party may act as if he has resigned.'
	This clause complies with the requirements of the HGCR Act 1996, section 108.
The *adjudication* W2.3(2)	'Within seven days of giving a notice of adjudication the Party referring the dispute to the *Adjudicator*
	• refers the dispute to the *Adjudicator*, • provides the *Adjudicator* with the information on which he relies, including any supporting documents and • provides a copy of the information he has provided to the *Adjudicator* • to the other Party.'
	This clause complies with the requirements of the HGCR Act 1996, section 108.
W2.3(5)	New requirement.
	'If a Party does not comply with any request or instruction of the *Adjudicator*, the *Adjudicator* may
	• continue the adjudication without that Party or document, and • make his decision based upon the information and evidence he has received.'
W2.3(10)	New requirement.
	'If the *Adjudicator* does not make his decision and notify it to the Parties within the time provided by this contract the Parties and the *Adjudicator* may agree to extend the period for making his decision. If they do not agree to an extension, either Party may act as if the *Adjudicator* has resigned.'
W2.3(11)	New requirement.
	'The *Adjudicator* may, within fourteen days of giving his decision to the Parties correct any clerical mistake or ambiguity.'
Review by the *tribunal* W2.4(1)	New requirement.
	'A Party does not refer any dispute under or in connection with this contract to the tribunal unless it has first been referred to the *Adjudicator* in accordance with this contract.'
	This clause makes it clear that all disputes must firstly go to *Adjudicator* as the first point of dispute resolution.

Dispute resolution procedure Option W2 *Continued*

W2.4(4)	New requirement.
	'If the tribunal is arbitration, the arbitration procedure, place where the arbitration is to be held and the method of choosing the arbitrator are those stated in the Contract Data.'
	The information required by this clause will need to be inserted into Contract Data part one.
W2.4(5)	New requirement.
	'A party does not call the *Adjudicator* as a witness in the *tribunal* proceedings.'

Index

Terms in *italics* are identified in Contract Data and defined terms have capital initial letters.
Page numbers in *italics* refer to diagrams or illustrations

acceleration
  programme, 36
acceptances
  programme, 24, 32–35
Accepted Programme
  costs, 21
  definition, 18–20
  failure to maintain, 38
  project management, 21
  purpose, 20–21
  resources, 20
  reviews, 37–38
  updating, 37–38
  use, 20–21
access dates, 22
access for *Employer*, 52
access of part of Site, 23–24
activity schedule, 38
actual progress
  programme, 30–31
additional testing, 53–54
adjudication
  dispute resolution, 75–76, *85*
  ECC3/ECC2 differences, 77–78
  information, 82–83
  options, *85*
  pre-HGCR 1996 Act, 75–76
  post-HGCR 1996 Act, 77
  process examples, 76
  tribunal, 76
Adjudicator, 82–84
  information judgement, 82–83
  involvement, 83
  records judgement, 82–83
ambiguities and inconsistencies
  Contractor's obligations, 42–43
  disputes, 74
  quality control, 42–43
  see also inconsistencies
amount due see assessing amount due
application for payment, 2–3
assessing amount due, 4–7
  details, 7
  included items, 5
  interest, 6
  time period, 6–7
  what is included, 5
assessment date for payment, 3–4

cashflow
  time control, 37
certification
  incentivisation, 59
  payment procedure, 7

quality control, 57–59
clause 14 type programme, *19*
Completion
  definition, 17
  payment procedure on Completion, 12–13
  payment procedure after Completion, 13
  time control, 35
Completion Date
  definition, 17
  programme inclusions, 22
concessions
  Defects, 57
conditions of contract
  payment procedure, 2–8
conflicts see disputes
construction methods/modes
  quality control, 44–45
contra proferentem rule, 74
Contract Data
  disputes, 69–70
contract date
  definition, 16
Contractor
  application for payment, 2–3
  programme, 36–37
Contractor's design
  quality control, 43–44
Contractor's employment
  termination, 60
Contractor's obligations
  ambiguities and inconsistencies, 42–43
  construction methods/modes, 44–45
  quality control, 41–47
    construction methods/modes, 44–45
    employees, 44
    management systems, 46–47
    supervision, 44
  setting out, 46
Contractor's plans, 33
Contractor's work
  order/timing, 25
control of time see time control
correction
  Defects, 55–56
    by Others, 59–60
costs
  Accepted Programme, 21
  disputes, 70–71

dates
  programme, 22–23, 24, 26–27
Defects
  concessions, 57
  correction, 55–56

Defects (*continued*)
   by Others, 59–60
   investigation, 53–54
   liability period, 56–57
   notification, 53–54
   work rejection, 55
Defects Certificate, 58–59
defects correction period definition, 57
defects date
   payment procedure after *defects date*, 13–14
   quality control, 57
defined terms, 16–18
definitions
   defects correction period, 57
   ECC2 *access date(s)*, 16–17
   ECC2 *possession date(s)*, 16–17
   Key date, 17
   Others, 49
   Planned Completion, 17–18
   programme, 18
   starting date, 16
   takeover by *Employer*, 35
   time control, 18–20
descriptors
   Works Information, 73
diaries, 38–39
   see also records
discrepancies *see* ambiguities and inconsistencies
dispute resolution, 63–87
   adjudication, 75–76
      options, 84–87, *85*
   NEC3 changes, 84–87
   Option W1 procedure, 84–85
   Option W2 procedure, 79–81, 86–87
   Star Chambers, 82
   tribunal, 78–81
   under ECC, 75–81
disputes, 63–87
   ambiguities and inconsistencies, 74
   costs, 70–71
   document interpretation, 65–71
   early warning, 71
   escalation, 82
   examples, 68, 72
   function/responsibility division, 72
   illegal actions, 74
   incidence reduction, 71–74
      function/responsibility division, 72
      Site Information, 73–74
      Works Information, 73–74
   origins, 64–71
      Contract Data, 69–70
      document interpretation, 65–71
      Site Information, 67–69
      Works Information, 66–67
   risk analysis perspective example, 68
   Site Information
      incidence reduction, 73–74
      origins, 67–69
   time effects, 70–71
   valuing changes, 71–72
   Works Information, 66–67

incidence reduction, 73–74
   origins, 66–67
documents
   disputes, 65–71

early warning of disputes, 71
ECC *see* Engineering and Construction Contract
employees
   quality control, 44, 59
   removal, 59
Employer
   programme, 37
   quality control, 49
Employer's part of works
   order/timing, 25
Employer's representatives
   quality control access, 52
   quality control role, 47–49
Employer's supply
   Plant and Materials, 49
   quality control, 49–50
enforcement
   quality control, 59–60
      low performance damages, 60
      removal of employees, 59
Engineering and Construction Contract (ECC)
   payment procedure, 1–14
      after Completion, 13
      after *defects date*, 13–14
      assessing amount due, 4–7
      assessment date, 3–4
      certification, 7
      on Completion, 12–13
      conditions of contract, 1–8
      Contractor's application for payment, 2–3
      ECC section 5/Option Y(UK)2 differences, 11–12
      HGCR 1996 Act, 8–10
      invoices, 7–8
      Option Y(UK)2 effects, 10–12
      representation, 8
Engineering and Construction Contract 2 (ECC2)
   access date(s) definition, 16–17
   interest, 6
   possession date definition, 16–17
Engineering and Construction Contract 3 (ECC3)
   HGCR Act 1996, 8–10
   interest, 6
   Option Y(UK)2, 9–10
escalation of disputes, 82

facilities and services
   quality control, 49
   see also Plant and Materials
failure to maintain
   Accepted Programme, 38
first assessment date, 3
first date for payment, 3
first programme submission, 31
float
   programme, 28–29
function/responsibility division
   disputes, 72

health and safety
    programme, 29–30
Housing Grants, Construction and Regeneration
    (HGCR) Act 1996
    ECC3, 8–10
    Option W2 dispute resolution procedure, 79–81

illegal actions
    disputes, 74
    Works Information, 74
incentivisation
    certificates/certification, 59
incidence reduction
    disputes, 71–74
included items
    assessments, 5
    programme, 21–31
inclusion notes
    programme, 22–31
inconsistencies
    Site Information, 74
    see also ambiguities and inconsistencies
information
    adjudication, 82–83
    included in programme, 21–32
inspections see test and inspections
interest
    ECC2/ECC3 differences, 6
    payment procedures, 6
investigation
    Defects, 53–54
invoices
    payment procedure, 7–8

Key Date
    definition, 17
    programme inclusions, 22

low performance damages, 60

making payment see payment procedure
Materials see Plant and Materials
method statements
    programme, 27

NEC2 clause changes
    quality control, 61–62
non-acceptance of programme, 33–34
non-compliance with Works Information, 33–34, 34
notification of Defects, 53–54

Option W1 procedure
    dispute resolution, 84–85
Option W2 procedure
    dispute resolution, 79–81, 86–87
    HGCR 1996 Act, 79–81
Option Y(UK)2
    ECC section 5 differences, 11–12
    ECC3, 9–10
    payment effects, 10–12
order/timing
    Contractor's work, 25

Employer's part of works, 25
Other contractors
    quality control, 49–50
Others
    definition, 49

payment procedure
    after Completion, 13
    after defects date, 13–14
    assessment
        amount due, 4–7
        date, 3–4
    certification, 7
    on Completion, 12–13
    conditions of contract, 1–8
    ECC, 1–14
    invoices, 7–8
    making payment, 8
    Option Y(UK)2 effects, 10–12
    periods for payment, 10–12
    programme, 5–6
people see employees
periods for payment procedure, 10–12
Planned Completion
    definition, 17–18
    programme inclusions, 22–23
Plant and Materials
    Employer's supply, 49
    programme, 24–25
    see also facilities and services
possession dates, 22
possession of part of Site, 23–24
pre-HGCR 1996 Act adjudication, 75–76
post-HGCR 1996 Act adjudication, 77
programme, 15–39
    acceleration, 36
    acceptance, 24, 32–35
        timing, 34
    activity schedule, 38
    actual progress, 30–31
    Contractor's plans, 33
    Contractor's work, 25
    dates, 22–23, 24, 26–27
    definition, 18
    Employer, 37
    Employer's part of works, 25
    examples, 19, 34, 34
    except first programme, 22
    first programme submission, 31
    float, 28–29
    health and safety, 29–30
    inclusions
        access of part of Site, 23–24
        Completion Date, 22
        Key Date, 22
        notes, 22–31
        Planned Completion, 22–23
        possession/access dates, 22
        possession/access of part of Site, 23–24
        starting date, 22
    information to be included, 21–32
    method statements, 27

programme (*continued*)
  non-acceptance
    reasons, 33–34
    timing, 34
  non-compliance with Works Information, 33–34, *34*
  order/timing of work, 25
  other aspects, 36–39
  other dates, 23
  other information, 30
  payment procedure, 5–6
  Plant and Materials, 24–25
  required information, 33
  resource statements, 28
  resubmission of unaccepted, 34
  revision frequency, 31–32
  special requirements, 30
  terminology, 16–18
  time effects example, *29*
  time risk allowances, 29
  timing
    acceptance, 34
    non-acceptance, 34
    of work, 25
  traditional clause 14 type, *19*
  updates, 31–32
  what is included, 21–31
  when submitted for acceptance, 18–20
  Works Information, 33–34, *34*
project management
  Accepted Programme, 21
Project Manager
  quality control, 48
  see also Employer's representatives

quality control, 40–62
  access for *Employer*, 52
  access for *Employer*'s representatives, 52
  additional testing, 53–54
  ambiguities and inconsistencies, 42–43
  certification, 57–59
  construction methods/modes, 44–45
  Contractor's design, 43–44
  Contractor's employment termination, 60
  Contractor's obligations, 41–47
  Defects, 55–57
    correction, 55–56
    defects correction period, 56–57
    investigation, 53–54
    notification, 53–54
  Defects Certificate, 58–59
  defects date, 57
  ECC elements, 41, *41*
  employees, 44
  Employer, 49
  Employer's representatives role, 47–49
  Employer's supply, 49–50
  enforcement, 59–60
    Contractor's employment termination, 60
    low performance damages, 60
    removal of employees, 59
  facilities and services, 49

  general, 52–54
  incentivisation, 59
  low performance damages, 60
  management systems, 46–47
  NEC2 clause changes, 61–62
  Other contractors, 49–50
  procedures, 54
  Project Manager's role, 48
  records, 58
  removal of employees, 59
  Subcontracting, 50–51
  supervision, 44
  Supervisor's role, 48–49
  test and inspections, 52–53
  work rejection, 55
  Works Information, 44

records
  Adjudicator's judgement, 82–83
  quality control, 58
  Supervisor, 38–39
removal of employees, 59
required information
  programme, 33
resource statements
  Accepted Programme, 20
  programme, 28
resubmission of unaccepted programme, 34
revision frequency
  programme, 31–32

scope
  works, 38, *38*
Secondary Option L, 17
sectional completion, 17
services *see* facilities and services
setting out
  Contractor's obligations, 46
site diaries/records, 38–39
Site Information
  dispute origins, 67–69
  dispute reduction, 73–74
  inconsistencies, 74
special requirements
  programme, 30
Star Chamber dispute resolution, 82
starting date
  definition, 16
  programme inclusions, 22
Subcontracting
  quality control, 50–51
submission
  first programme, 31
supervision
  quality control, 44
Supervisor
  diaries/records, 38–39
  quality control, 48–49
  see also Employer's representatives

takeover by *Employer*, 35
termination

Contractor's employment, 60
terminology
    programme, 16–18
test and inspections, 52–53
time control, 15–39
    acceleration, 36
    Accepted Programme definition, 18–20
    cashflow, 37
    Completion, 35
    programme definition, 18
    takeover by *Employer*, 35
    terminology, 16–18
time effects
    disputes, 70–71
    programme example, *29*
time period assessment, 6–7
time risk allowances
    programme, 29
timing
    programme acceptance, 34
    programme non-acceptance, 34
traditional clause 14 type programme, *19*

tribunal, 76, 78–81

updates
    programme, 31–32

valuing changes
    disputes, 71–72

work rejection
    quality control, 55
works
    order/timing, 25
    scope, 38, *38*
Works Information
    conflicts, 73–74
    descriptors, 73
    disputes
        origins, 66–67
        reduction, 73–74
    illegal actions, 74
    quality control, 44
    special requirements, 30